Boundedness and Self-Organized Semantics:
Theory and Applications

Maria K. Koleva
Bulgarian Academy of Sciences, Bulgaria

Managing Director:	Lindsay Johnston
Editorial Director:	Joel Gamon
Book Production Manager:	Jennifer Romanchak
Publishing Systems Analyst:	Adrienne Freeland
Development Editor:	Hannah Abelbeck
Assistant Acquisitions Editor:	Kayla Wolfe
Typesetter:	Henry Ulrich
Cover Design:	Nick Newcomer

Published in the United States of America by
Information Science Reference (an imprint of IGI Global)
701 E. Chocolate Avenue
Hershey PA 17033
Tel: 717-533-8845
Fax: 717-533-8661
E-mail: cust@igi-global.com
Web site: http://www.igi-global.com

Library of Congress Cataloging-in-Publication Data

Koleva, Maria K., 1958-
  Boundedness and self-organized semantics: theory and applications / by Maria K. Koleva.
    pages cm
  Includes bibliographical references and index.
  Summary: "This book enhances the understanding of the theoretical framework and leading principles of boundedness, aiming to bridge the gap between biology, artificial intelligence, and physics"-- Provided by publisher.
  ISBN 978-1-4666-2202-9 (hardcover) -- ISBN (invalid) 978-1-4666-2203-6 (ebook) -- ISBN (invalid) 978-1-4666-2204-3 (print & perpetual access) 1. Self-organizing systems. 2. Functions of bounded variation. 3. Semantics (Philosophy) I. Title.
  Q325.K647 2013
  501'.17--dc23
                    2012019290

British Cataloguing in Publication Data
A Cataloguing in Publication record for this book is available from the British Library.

All work contributed to this book is new, previously-unpublished material. The views expressed in this book are those of the authors, but not necessarily of the publisher.

Table of Contents

Section 2
Boundedness: Properties of Self-Organized Semantics

Preface

Complex systems is a new field of science that aims to provide answers to the questions: how parts of a system give rise to a variety of its collective behaviors, and how the system interacts with its environment. It is easy to enumerate examples of complex systems. These are, for example, the social systems whose constructive elements are the people; the brain as a biological system is composed out of neurons; molecules are formed out of atoms; the weather is formed out of air flaws. This new field of study of complex systems cuts across all traditional disciplines of science, as well as engineering, management, and medicine.

The intensive empirical examination that was going on in the last decades displays the remarkable enigma of their behavior: the highly specific for each complex system properties persistently coexist with certain universal, shared by each of them ones. Thus on the one hand, they all share the same characteristics such as power law distribution and sensitivity to environmental variations, for example; one the other hand, each system has its unique "face," i.e. we can distinguish between an earthquake and heartbeat of a mammal. What makes the study of this coexistence so important is the enormous diversity of systems where it has been established. In order to get an idea about this vast ubiquity let us present a brief list of such phenomena: earthquakes, traffic noise, heartbeat of mammals, public opinion, currency exchange rate, electrical current, chemical reactions, weather, ant colonies, DNA sequences, telecommunications, etc.

But the greatest mystery enshrined in the behavior of complex systems is that both intelligent and non-intelligent systems belong to the same class: thus a Beethoven symphony, a product of a genius mind, and the traffic noise which, though being also a product of human activity, but un-intelligent in its behavior, share the same type of power spectrum. Another example is the semantics of human languages: in the year 1935 the linguist G. K. Zipf established that, given some corpus of natural languages, the frequency of any word is inversely proportional to its rank in the frequency table. Thus the most frequent word will occur approximately twice often as the second most frequent word, three times as often as the third most frequent word, etc. Put in other words, the Zipf law ignores any semantic meaning and thus it seems to sweep out the difference between mind activity and random sequences of letters. Thus we come to the fundamental problem of the present book: what makes a complex system "intelligent" and why it should share such "indifferent" to the intelligence properties?

The affiliation of the apparently intelligent systems such as human languages and music to the same class as earthquakes and a variety of other natural phenomena, suggests in a straightforward way that the intelligent behavior is embedded in natural processes. Thus, the opposition between intelligence as highly specific activity and the fact that it inherently belongs to a universal class of natural phenomena raises the major question whether it is likely to expect defining a criterion able to distinguish the specificity of a system from another one along with affiliating each of them to the same class of complex systems?! If exists, such criterion would gain an enormous ubiquity and power from its relevance to hierarchical-like organization and functioning of a wide class of complex systems. If so, it is to be expected that, like the semantics of human languages, their hierarchical-like self-organization would be non-extensive as well. Then, the non-extensivity of semantic-like intelligent hierarchical-like structures would constitute their major distinctive characteristic, namely their irreducibility to a Turing machine.

The above considerations substantiate the major goal of the present book: we put forward a general principle, named by us boundedness, that governs both the organization and functioning of complex systems so that to allow: (i) explaining how and why both intelligent and non-intelligent complex systems share the same type universality; (ii) unambiguous discrimination between intelligent and non-intelligent complex systems: (iii) providing the general frame for non-extensive hierarchical structuring of the information. The distinctive feature of that non-extensivity is its organization and implementation in a semantic-like manner. It is important to stress that a crucial notion of our concept is the association of a physical implement to the semantic meaning so that the latter to be irreducible to the sequence of symbols that constitute it.

A fundamental difficulty for defining such principle is that we are bound to leave the paradigm about expressing it in the form of a law, either deterministic or stochastic. In order to make this surprising statement clearer let us remind what the notion of law asserts: a law quantifies specific relation between certain, specific for the event, variables. The most important property of any law is its time-translational invariance, i.e. the law remains the same on reoccurrence of the event in any time. However, what is tacitly presupposed is that, on reoccurrence, the environmental conditions for setting the law also reoccur. In turn this requirement bounds the law to be local. This viewpoint is embedded in the widespread probabilistic approach for modeling the environmental impact and the corresponding response: the variability of the environment is presented through probabilistic distribution and after each session of impact, the system is supposed to return to the state of equilibrium which is global attractor for all initial conditions. Up-to-date, the validity of this view seems so unshakable that it is wildly applied for modeling almost everything in Nature and human activity. However, this setting is completely inappropriate for modeling the behavior of complex systems since neither the intelligent activity is subject to any probabilistic distribution, nor the homeostasis of the living organisms is reducible to a single equilibrium state.

The following two empirical observations turn crucial for supporting the rejection of the probabilistic approach and for defining an appropriate general theory of the behavior of complex system:

1. The intelligent type behavior is always associated with permutation sensitivity of its constituents - indeed changing the order of letters in a word and the order of the words in a sentence is crucial for their meaning; the same is with the music – any change of the notes succession could turn the music into noise;

2. Complex systems, both intelligent and non-intelligent, are characterized by robustness to environmental variations if only the later do not exceed certain, specific for each system, thresholds. Yet, not only the amplitude of the environmental variations is subject to restraint but the increments of their succession are subject to restraint as well. Perhaps the most delicious example of the phenomenon of imposing this restraint comes from cooking: our daily experience tells us that the best meal is cooked in a slow oven. Another common example comes from metallurgy: since ancient times it is proven that the finest and strongest pieces are product of masterpiece recipes the goal of which

is to regulate, i.e. to put bounds on the speed of variations of temperature, concentrations etc., in the process of crafting the product.

The deduction from the second of the above empirical observations suggests that both variations of the environmental changes and the response of a system are to be bounded so that the system to stay long-term stable. Indeed, our daily experience teaches us that on exceeding certain thresholds of the impact and/or its increment, every system would fall apart.

The inference from the above considerations has its understanding in the formulation of the general principle of boundedness as follows: 1) a mild assumption of boundedness on the local (spatial and temporal) accumulation of matter/energy at any level of matter organization and 2) boundedness of the rate of exchange of such an accumulation with the environment. Also, it should be taken into account the notion of unconditional release from any predetermination of the local response as an essential ingredient of the concept.

Now, it becomes evident why the boundedness is rather a principle of organization than a law. Indeed, while a law quantifies the local response requiring its exact reoccurrence in any moment, its release gives enough power to the concept of boundedness to quantify the notion of homeostasis as specific self-organized response to an ever-changing environment so that to keep on the long-term stable functioning and specific order of a system. Then, the universal properties of the complex systems appear as necessary ingredients for maintaining homeostasis in an ever-changing environment.

The fundamental difference between the concept of boundedness and the probabilistic approach becomes most evident when considering the time-translational invariance of the homeostasis. It is grounded on the proof that the power spectrum of any bounded time series is additively decomposed into a specific discrete band and universal, noise one. The discrete band characterizes emergent (self-organized) pattern while the exclusive property of the noise band is its invariance to the environmental statistics. These properties of the power spectrum ensure the time-translational invariance of the discrete band (i.e., of the self-organized pattern) by means of maintaining constant accuracy of its reoccurrence in any time. Thus, not only the emergent pattern gains the major characteristics of homeostasis, namely it is self-organized object that remains the same in an ever-changing environment, but its characteristics are irreducible to any weighted average as it would be with the probabilistic approach.

The differences between the probabilistic approaches and the concept of boundedness are conceptual, i.e. it is not that the concept of boundedness renders the probabilistic approaches irrelevant but it is rather that it renders their understanding and utilization to acquire novel comprehending. Chapter 5 demonstrates how crucially the role of probabilistic approach depends on the current views on the subject-matter. It demonstrates that we encounter two fundamentally different settings for the probabilistic approach: the traditional one considered above and a new one which comes in result of considerations about the crucial role our understanding plays in substantiating the notion of "measurement." Thus: (i) the traditional approach assumes primary importance to the accuracy of one of the major universal characteristic of the complex systems, namely power law distributions, that remains in a stark confrontation with time-translational invariance and empirically observed stability of the complex system; (ii) our approach considers the power law distributions as a result of "destructive" measurements which, however, retains the exclusive property to serve as a universal criterion for hierarchical structuring under boundedness. In turn, the well developed techniques of the probabilistic approach acquire new role that helps establishing hierarchical structuring.

The self-organization of the homeostasis in spatio-temporal patterns poses the question how to distinguish them from one another and if so how to quantify the difference. The traditional approach involves as a measure the energy/matter involved and exchanged in the process of its formation. Despite its undisputable relevance, such measure is indifferent to the "efforts" made by the system to maintain the homeostasis. That is why we put forward the assumption that the discrete band of the power spectrum is an appropriate candidate for characterization of homeostasis because of its two-fold meaning: (i) on the one hand it serves as a measure for energy/matter involved in the pattern and as a necessary element for sustaining its homeostasis; (ii) at the same time it serves as characteristics of its structure - the latter is embedded in the specific order and in the ratios established between the member frequencies. The decisive step further is taking into account the permutation sensitivity of the member frequencies which makes the discrete band of the power spectrum the best candidate for information symbol. It is worth noting the fundamental difference of the present definition of information compared to the traditional one: while the latter assigns a number (probability) that is a product of our understanding for the system and Nature (i.e., it is a product of an external mind), the association of the information with the discrete band in the power spectrum makes this information autonomous i.e., independent of our understanding.

Once again, recall the two-fold way of characterization of any information symbol by means of reading the power spectrum: (i) through energy/matter involved and exchanged for either organizing the pattern and it's sustaining; (ii) classifying the structure of the pattern through specific order of the member frequencies. The importance of the two-fold characterization is that it allows executing non-recursive computing in finite number of steps, namely: information symbols whose discrete bands comprise algorithmically non-computable numbers (irrational ones) are now physically reachable by finite number of steps since the matter/energy exchange for covering the distance between the symbols is finite.

The next difference between the traditional information theory and complex systems under boundedness is given by the following consideration: due to the boundeness of the rates, any given information symbol is reachable from its closest neighbors only. Thus the next and the previous symbol in a sequence could be "foreseen" on the grounds of the knowledge about the current unit only, i.e. such circuit performs as an "Oracle." To compare, according to the traditional information theory each letter can follow or precede any other one from the alphabet and thus it is impossible not only "foreseeing" the next letter, but it is impossible to judge *a priori* whether a given sequence is a message (i.e. it comprises information) or it is a random sequence of symbols. However, though "foreseeing" the next symbol is important in many aspects, still it is not self-organized semantics.

What is decisive for natural processes to execute in a semantic-like manner, both spontaneously or driven, is the opportunity for physical implementation of a "space bar." For making us clearer let us briefly consider general structure of the state space of an intelligent complex system: 1) under the assumption of the hypothesis of boundedness the state space of an open system is partitioned into basins-of-attraction and the trajectories form a dense transitive set of orbits; each basin is characterized by its own spatio-temporal pattern which is intra-basin invariant. The motion in the state space is implemented by restricting transitions from a point on a trajectory to being those to nearest neighbors only; i.e., transitions from any point on a trajectory must pass through an adjacent state (one can think of a latticised state space); 2) the state space has an accumulation point, namely it is the point which is simultaneously tangent to all basins-of-attraction. The importance of the presence of an accumulation point is that it plays the role of a "space bar." The interplay between the orbital-type motion and the presence of a "space bar" permits one-to-one correspondence between words and orbits so that each word is separated from its neighbors by "space bar." The immense importance of the physical implementation of the "space bar" is that it makes the process of decoding information autonomous from the "external mind."

The decisive step forward is the association of the meaning of each "word," "sentence" (semantic unit) with the performance of the "engine" constituted by the corresponding orbit. The power of this association lies in the functional irreversibility of any Carnot engine: in one direction it works as a pump and in the opposite it performs as a refrigerator; then, the function irreversibility of the engines implements permutation sensitivity of semantic units. The exclusive role of the two-fold presentation of a semantic unit (word, sentence) is that this way it simultaneously implements not only non-recursive computing but performs as implement for building non-extensive hierarchical-like organization of semantic-like type.

Now the irreducibility of the semantic-like behavior of intelligent complex systems to a Turing machine turns apparent: indeed, the universality of Turing machines does not allow any autonomous discrimination between orbital and any non-orbital motion since the universality implies every string admissible. Consequently, for Turing machines, the semantics of any string of symbols is revealed only by means of a decoder set by an "external mind." Another fundamental difference between them is that while the Turing machines proceed only through linear processes and thus process only natural numbers, the two-fold presentation of information symbols and semantic units of intelligent complex systems allows processing of non-recursive numbers in finite number of steps.

The general conclusions that can be drawn are the following: (i) the greatest value of a Turing machine is that it can execute any recursive algorithm; however, this comes at the expense of non-autonomous comprehending of the obtained output. On the contrary, the ability of the introduced by us non-recursive computing to execute any algorithm is constrained by the boundedness which precludes transitions which exceed thresholds; at the same time, it substantiates the fundamental value of that approach – autonomous comprehending. Thus, the Turing machine and the non-recursive computing appear rather as counterparts than as opponents.

Only a brief outline of the major characteristics of our approach underscores the exclusive property of the proposed approach to be the intelligent-like way of self-sustaining the functionality of the multi-layer non-extensive hierarchy by means of governing the intra- and inter-level homeostasis through rules that are organized in semantic-like manner.

The goal of this Preface is to make a brief introduction to the power and utility of the concept of boundedness. In it some of the major themes have already been mentioned. But this new theory is rich enough that allows other important topics to be discussed in the book. For example, one of the most exciting ones is the issue whether the concept of boundedness supports the Second Law. The problem is that the concept of boundedness meets fundamental difficulties with its most widespread formulation, namely the one that asserts: the equilibrium is the state of maximum entropy. We start revealing the confrontation with pointing out that the traditional notion of entropy encounters fundamental difficulties when applied to structured systems since any order and/or structure implies by definition that the corresponding state is *not* the state of maximum entropy. Alongside, the replacement of the idea of equilibrium with that of homeostasis and the assumption that not a single basin-of-attraction can be a global attractor substantially contribute to the confrontation. On the other hand, the idea of implementation of semantic meaning through a specific "engine" implies that semantics is "hard work" supplied by the environment. Thus it is important to quantify the relation between the "work" produced by the "engine" and the energy/matter involved in the corresponding process. So, we face the question whether the concept of boundedness yields a *perpetuum mobile* and if not, how and why it happens; is the ban over *perpetuum mobile* a general result and if so how the concept of boundedness modifies the idea of Second Law.

A critical comparison between the traditional algorithmic approach and the semantic-like one is made in Chapter 10. One of the major differences between them turns out to be that whilst the semantic-like approach permits autonomous discrimination between "true" and "false" statement by an intelligent complex system, the traditional algorithmic theory which does not allow any autonomous discrimination between a "true" and a "false" statement. Yet, the differences between both approaches highlight that they are rather counterparts than opponents. Thus, while we can suppose any logical statement by means of the traditional algorithmic approach, the semantic-like one selects only physically substantiable statements.

Acknowledgment

The author would like to acknowledge the help of Prof. Donald L. Bennett and Prof. Holger Beck Nielsen from Niels Bohr Institute in Copenhagen for the useful discussions during my visit to Niels Bohr Institute in 2009 without which the present book could not have been satisfactorily completed.

Deep appreciation and gratitude is due to my mother and my brother for the continuous moral support for this years-long project.

Section 1
Boundedness and Stability:
Characteristics of Homeostasis

INTRODUCTION

Boundedness and Stability: Characteristics of Homeostasis

This book is an attempt to explain the enigmatic coexistence of universal and specific properties of the complex systems in a unified way by establishing the concept of boundedness. The enormous diversity of complex systems precludes the possibility that all detailed observations can be condensed into a small number of mathematical equations, similar to the fundamental laws of physics. Moreover, this huge diversity strongly suggests that any successful explanation of the universal characteristics should be insensitive to the mathematical description of their physical nature. Additionally, these considerations suggest that the universality is to be insensitive to the details of environmental variations and the statistics of the internal fluctuations of any system. The universal properties also should be insensitive to whether the local response is linear or non-linear. The unified explanation of both universal and specific properties constitutes one of the greatest challenges posed by the empirical observation that while the universal properties are parameter-free (e.g. power laws), the specific ones are characterized by self-organized patterns each of which has its own structure. Another challenge to any successful explanation is to prove that each of these properties along with their coexistence admit time-translational invariance and thus provide stable long-term behavior of the complex systems.

The section defines the concept of boundedness as boundedness on the local (spatial and temporal) accumulation of matter/energy at any level of matter organization and 2) boundedness of the rate of exchange of such an accumulation with the environment. An important ingredient of this concept is that the characterization of the local impact and the response to it is released from any predetermination. To compare, the widespread statistical approach presupposes the environmental impact and the corresponding response to obey the same predetermined distribution and bounds the local response to start and end at equilibrium; then, for any macroscopic variable stands a weighted average. However, boundedness introduces non-physical correlations among distant responses and thus makes the idea of equilibrium irrelevant. That is why we replace the idea of equilibrium with the idea of homeostasis which asserts that any non-specified, yet bounded response, is separable to a specific part given by a self-organized pattern, and a parameter-free universal part. Thus the self-organized pattern stands for the specific non-local response of a system whose major property is robustness to small environmental variations. Since this is exactly the definition of homeostasis as one can find in the textbooks, we will use this term further in order to highlight its fundamental difference from the notion of equilibrium.

What makes this task highly non-trivial is the controversy in mathematical description of homeostasis: on the one hand, the robustness of homeostasis renders it invariant to a bounded set of environmental variations; put it in mathematical terms its characteristics must be intra-basin invariants and thus expressed in metric-free forms; on the other hand, the boundedness of the rates always defines specific metrics so that in every moment a system "knows" where it is and where it is allowed to go. So, the question is how to define a metric-free expression in a metric space?!

The goal of this part of the book is to present with the sufficient mathematical rigor a systematic theory for characterization of the coexistence between the universal and the specific properties of complex systems. All of the above items will be considered properly along with others among which is the issue how a system "feels" the thresholds so that to make a "U-turn" instead of bumping transversely into it?! Decisive for the major goal of the book is the question about the discrimination of the intelligent systems from the non-intelligent ones. On the grounds of the principles developed in the first five chapters we will be able firmly to assert that the pre-condition for emergence of the intelligent-like hierarchical organization is a system to be subject to morphogenesis; the key point in the description of these systems is the proof that the boundedness of the rates appears automatically as a result of the application of the concept of boundedness to the quantum-mechanical level. What is important to say is that the rates of admissible transitions explicitly depend on certain quantum-level characteristics and thus substantiate an inter-level feedback.

Thus, the hierarchical-like order appears as generic implement for sustaining homeostasis by means of substantiating intra- and inter-level interactions through specific feedbacks which operate in a semantic-like manner. Therefore the diversity of properties is reached by diversification of the hierarchical structuring. It should be stressed that this setting makes our approach fundamentally different from the traditional reductionist philosophy of physics. Indeed, the latter reduces the complexity to separate levels of organizations (elementary particles, atoms and molecules, large systems, biological systems) each of which is considered at specific pre-determined external constraints so that the hierarchical order goes only bottom up i.e., from elementary particles to living organisms. These assumptions constitute its fundamental difference from the concept of boundedness which, on the contrary, sets hierarchical self-organization at non-specified external constraints so that its maintaining goes both bottom up and top down by means of loop-like inter-level feedbacks that serves as constraints imposed by one level to the nearest (both upper and lower) ones. Further, on the contrary to the reductionist approach which adopts the idea of dimension expansion as a tool for reaching diversity of properties, the diversification under boundedness happens through hierarchical super structuring in constraint dimensionality.

A decisive test for the relevance of the concept of boundedness is the incorporation of the universality of the power laws in its frame– on the one hand, an essential ingredient of the concept of boundedness is the release from any pre-determination of the impact and response which mathematically is expressed as insensitivity of certain characteristics to the statistics of the impacts and responses; on the other hand, the huge amount of empirical observations displays the same type power law distributions for the response of the widest variety of complex systems. Chapters 5 and 6 demonstrate that not only the power laws naturally appear in the concept of boundedness but it will be demonstrated that their major property, i.e. to be parameter-free, peacefully coexists with the specificity of the corresponding spatio-temporal pattern which characterizes homeostasis. Moreover, it will become evident that the power laws and the other universal characteristics appear as characteristics of the inter-level feedbacks of a multi-level hierarchical structure. Yet, the most non-trivial property of all inter-level feedbacks is its participation in a recursive way to the discrete band of a power spectrum, i.e. to an information symbol.

Chapter 1
Time Series Invariants under Boundedness:
Existence

ABSTRACT

It is proven that every zero-mean bounded irregular sequence (BIS) has three invariants, i.e. characteristics which stay the same when the environmental statistics changes. The existence of such invariants answers the question how far they ensure certainty of the obtained knowledge and the range of predictability of stable complex systems behavior in a positive way. The certainty of our knowledge is put to test by the lack of global rule for response makes impossible to adjust a priori the corresponding recording equipment to a long run. Then, it is to be expected that the recorded time series does not match the corresponding signal in a uniform way since the record is subject to local distortion which is generally non-linear and acts non-homogeneously on the recording. In turn, this poses the fundamental question whether it is ever possible to establish and/or predict the properties and the future behavior of the complex systems.

INTRODUCTION

One of the most exciting properties of the complex systems is that even though each of them responds in a highly specific manner to the tiniest changes of the environment, most of them remain remarkably stable in the following sense: a stable complex system exhibits specific steady pattern of behavior whose properties are insensitive to the time window of any observation. A remarkable general property of that stability is that though the response is highly specific, all stable complex

DOI: 10.4018/978-1-4666-2202-9.ch001

systems share certain universal properties. This well established empirical fact prompts us to put forward the idea that the response of stable complex systems is decomposable to a specific steady part and universal noise one. For reasons that would become clearer later we suggest to call the specific part of the response homeostasis and the non-specific one to call fluctuations. An exclusive property of stable complex systems, proven in Chapter 2, is that their response, represented by corresponding power spectrum, resolves explicit additive separation to a specific pattern and a universal noise band. Taking for granted the proof of this claim until Chapter 2, now we focus our attention on establishing those exclusive properties of a noise band which commence from stability of the corresponding complex system alone.

The derivation of the time-series invariants follows Chapter 1 of our previous work (Koleva, 2005).

The first step in this detailed presentation is to establish a general frame for description of the enormous diversity of irregular variations which characterizes the response of the stable complex systems. These irregularities constitute a significant part of our daily life: traffic noise, heartbeat, public opinion, currency exchange rate, electrical current, chemical reactions - they all permanently and irregularly vary in space and in time. The remarkable property of this behavior is that, though wildly varying, each of these systems "keeps" its "individuality" intact - at every moment we can distinguish without any moment of hesitation between a heartbeat in a mammal and a daily record of currency exchange rate. All these diverse phenomena can be conceptualized by a precise definition of a fluctuation that can be expressed as: a fluctuation is any deviation from the steady pattern which serves as "identity card" of the corresponding system. The advantage of this definition is that it renders the notion of a fluctuation independent of the enormous diversity of the "driving "forces": human emotions, varying interests in economic systems, and physical

interactions. The existence of such definition opens the door to a powerful general study whose outcomes are relevant to each and every system regardless to its origin and particularities. In turn, this makes the subject an indispensable part of the fundamental science. Indeed, for a long time an abstract definition of the fluctuations does exist - irregular deviations from the average that leaves the system stable; the average stays steady while the amplitude and the frequency of the deviations are not precisely predictable. The application of this definition in the analysis of the behavior of a fluctuating system leaves only one possible approach: the systems` response can only be matched through the properties of an irregular time series. This general understanding involves a lot of efforts and ingenuity in order to study characteristics such as the frequency and size of the fluctuations, first exit time etc. The overall thread that unites the enormous number of studies dealing with this type of analyses is that they are specific to the statistics of the fluctuations. That is why the mainstream research has been set on various statistical approaches and the associated classifications according to appropriate parameters.

However, little or no attention has been paid to the issue about how a general phenomenon such as long-term stability of the complex systems do exist. The question is how system regulates the characteristics of its fluctuations in order to stay stable, i.e. how it regulates its characteristics so that its response to retain a steady pattern whose characteristics are insensitive to the time window of any observation. Again our daily experience helps to adjust our definition to the new constraint - stability exists only when the fluctuations and the rate of their development do not exceed certain specific for the system thresholds. Otherwise the system undergoes either qualitative changes or collapses. Furthermore, the "distance" to the thresholds can be subjective as in the case of our emotions. Yet, we assert that any "distance" to the thresholds of stability is always finite. This understanding of existence of only limited ranges of distances to thresholds comes as a generalization of our daily experience and says that all the objects in Universe are created by involving finite amount of energy and/or matter and that their behavior in an ever-changing environment stays stable if only the energy/matter involved and exchanged in the process is kept permanently bounded within specific margins. The concept of boundedness has been put forward in our paper (Koleva & Covachev, 2001).

BACKGROUND

The above statement substantiates the major part of the introduced by us concept of boundedness which we put forward for explaining the behavior of complex systems. An indispensable part of our concept is the release of the local response from any pre-determination. This implies that the local response is specified by the current impact and the current state of the system only and that the impact is arbitrary, but bounded. Note that thus the response is indeed local, i.e. free from necessity of any return to equilibrium as it would be with the statistical approach; the locality implies also that it could vary from being linear under given impact to non-linear under another impact. To compare, any *a priori* imposed distribution pre-determines the properties of the response in a long-term run. Indeed, while the locality of the response is achieved through assuming returns to the equilibrium after each session of impact, certain properties of the response, e.g. to be linear or non-linear at every impact, are globally pre-determined by *a priori* setting of the statistical distribution of the environmental variations. Yet, it seems that the withdrawal from the traditional approach makes our task enormously difficult since we lose the major implement for providing long-term predictions. Indeed, the highest popularity of the statistical approach comes from the prospect for explicit long-term predictions

encapsulated in a compact form, implemented by the *a priori* setting of the environmental statistics. Unfortunately, this approach is vulnerable to every tiny uncertainty in measurements since the error is gradually accumulating in a long-run. In turn the compact form of prediction comes at the expense of demand for equipment that has the quality to execute measurements with absolute precision. The general susceptibility of long-term predictions from the impact of environmental statistics defines the fundamental importance of the task for looking for another general approach to the behavior of complex systems and its predictability.

The major goal of the present chapter is to prove that every zero-mean bounded irregular sequence (BIS) has three invariants, i.e. characteristics which stay the same when the environmental statistics changes. The existence of such invariants and their proper understanding will help answering a lot of questions among which are the questions how far they ensure certainty of the obtained knowledge and the range of predictability of the complex systems behavior. The certainty of our knowledge is put to test by the irremovable uncertainty in the recording of the fluctuations in time series. This uncertainty appears because the lack of global rule for response makes impossible to adjust *a priori* the corresponding recording equipment to a long-term run. Then, it is to be expected that the recorded time series does not match the corresponding signal in a uniform way since the record is subject to local distortion which is generally non-linear and acts non-homogeneously on the recording. In turn, this poses the fundamental question whether it is ever possible to establish and/or predict the properties and the future behavior of the complex systems. The positive answer to this question comes from the proof that a BIS is invariant to "coarse-graining" which is a mathematical operation that involves non-linear operations like local averaging, local amplification and/or local damping, filtering, resolution effects etc., and which acts non-linearly and non-homogeneously on the original BIS. Thus the time series invariants expressed in an invariant under the statistics of environmental impacts and under the unavoidable coarse-graining of the recording compact form appear as grounds for providing certainty of our knowledge about the behavior of complex systems and its predictability. Moreover, one of the time series invariants, i.e. power spectrum, serves as an implement for establishing the tiniest development of any change in the homeostasis though it is not yet "foreseeable" on monitoring the corresponding time series.

Next we will prove that the static and the dynamical boundedness (boundedness of rates) indeed make every BIS to display time-series invariants. We will prove that there are 3 characteristics of a BIS which are insensitive to the statistics of its members. They are: the shape of the autocorrelation function, the shape of the power spectrum, and the shape of the K-entropy. These features are accompanied by the universality of the population of an attractor. It is worth noting that the existence of the time-series invariants is an exclusive property of boundedness which is not shared by any *unbounded* counterpart. Indeed, the shape of a power spectrum of an *unbounded* time series explicitly depends on the statistics of the members and the length of the corresponding time series. The fundamental importance of the existence of time-series invariants will be best revealed in the next Chapter where we prove the additive separability of the power spectrum of a BIS to a specific discrete band (which represents the corresponding homeostatic pattern) and a noise band of universal shape. The value of this decomposition is that it renders constant accuracy of both occurrence and recording of a given homeostatic pattern. The major move ahead, developed at the end of Chapter 2, is that on the grounds of this separability we are able to work out early warning signs for any change in the homeostasis.

Along with the matter of predictability, the existence and the properties of the time series invariants opens the door to a metric-free measure of information through its association with a robust self-organized pattern (homeostasis). Note that in the traditional information theory the information is associated with certain number (probability) assigned to the current state of the system according to our current "views" on the system, i.e. the information is subjective and a product of an "external mind".

Yet, we postpone to Chapter 2 the considerations how the time-series invariants are implemented for elaborating a criterion for predictability. The goal of the present chapter is to prove the existence of the time series invariants and to establish their explicit compact form.

Let us start our consideration by the most general claim: we assert that all BIS of infinite length share the same asymptotic properties regardless to when the fluctuations commence and the details of their development. Alone the assumption of boundedness allows us to understand common properties of systems as diverse as quasar pulsations, financial time series, heartbeat and fluctuations in electric current.

The first step in presenting this complex behavior is to specify the notion of a BIS. An example of a BIS is the record of daily variations of the currency exchange rates. The precise definition of a BIS is the following one: any stochastic sequence the terms of which succeed in arbitrary order but the size of the terms is limited to be within given margins. The terms correspond to the fluctuations and the margins are dictated by the thresholds of stability.

The length of a BIS is determined by the time interval during which the system stays stable. In this chapter we consider only BIS of infinite length. Applied to the real world this means that the system in question stays stable permanently. This property of a system puzzles our normal reasoning and intuition by posing the question: what will happen when a large enough fluctuation reaches the thresholds of stability? On the one hand, if the fluctuation "bumps" into the threshold transversely, it destabilizes the system. So, depending on the "bumping" angle, the further evolution is either terminated due to instability or requires a specific response in order to stabilize the system. However, the stabilization is always temporary because one of the next "bumps" will most likely cause the collapse of the system. Hence, in order to provide an arbitrarily long-term stability, the fluctuations must make a "U-turn" at their approach to the thresholds of stability. The far-reaching non-triviality of this issue lies in the necessity of reconciliation of two seemingly opposing requirements: on the one hand, in order to keep homeostasis intact, no physical process must be involved at a U-turn; on the other hand a fluctuation must "feel" the thresholds by some means so that accordingly to execute an U-turn. This is a key question in the present book. The mechanism behind the U-turn sets a new role of the stretching and folding mechanism; to remind that the latter brings about the deterministic chaos in simple dynamical systems.

Since the general mechanism for the "U-turns" requires only very weak conditions about the properties of the fluctuations and their origin, it is reasonable to postulate the relevance of the U-turn mechanism for a broad spectrum of systems. Although the specification of these conditions is postponed until Chapter 5, the idea associated with the notion of "U-turn" is that the development of any given fluctuation does not initiate a process behind any specific response of the system. Hence, the sequence is scale-free, i.e. all the time scales contribute uniformly to the properties of the system.

The properties of BIS which we will establish in the present Chapter and the next one help to resolve a fundamental problem faced by all modern laboratories: that is the effect of inevitable distortions of the recorded data introduced by the measuring apparatus. Despite sensitivity and ingenuity of the latest experimental equipment, its complexity makes the obtained results rather distorted by processes ranging from local averaging, to local amplification and/or local damping, filtering, resolution effects etc., i.e. these are

non-linear operations that act non-homogeneously on the original BIS. This fact poses the question whether we can ever hope to separate the specific characteristics of any system from the "noise" if we are not able to "record" any signal precisely?! One of the greatest advantages of our approach is that this problem does not exist for BIS. Its successful resolution is grounded on the exclusive property of the time-series invariants of a BIS to be insensitive to any distortion of their members if only it leaves their boundedness intact. Hereafter we denote the mathematical operation that acts non-linearly and non-homogeneously on members of a BIS, under the mild constraint of preserving boundedness of each and every of them, as operation of coarse-graining.

This problem outlines the fundamental importance of our statement that BIS have certain common properties and makes the goal of our book two-fold - by establishing the general properties of the fluctuations we must outline the route to the specific ones.

Now we can formulate the most fundamental property of the BIS of infinite length: each BIS remains bounded and scale-free upon coarse-graining.

TRANSFORMATIONAL AND SCALING INVARIANCE

Now we shall illustrate how the coarse-graining operates by means of one of its "tools", namely local averaging. Given a BIS that may be either discrete, (i.e. its members are x_i placed at time points t_i) or continuous, (i.e. to be a bounded stochastic function $x(t)$). For the sake of convenience we will develop what follows in terms of functions.

The local average is defined in the window $[T_1, T_2]$:

$$\overline{x}_{T_1 T_2} = \frac{1}{(T_2 - T_1)} \int_{T_1}^{T_2} x(t)dt \qquad (1.1)$$

Recalling that the expectation value of an arbitrary sequence is defined as:

$$\langle x \rangle = \lim_{T \to \infty} \frac{1}{T} \int_0^T x(t)dt \qquad (1.2)$$

The question of how "close" the local average is to the expectation value arises naturally. Obviously, when the window $[T_1, T_2]$ goes to infinity, it is to be expected that the local average converges to the expectation value. Suppose that for the BIS the expectation value does exist and is finite. This will be proved later. Now, along with this question we address the issue what happens when the window is finite. The latter problem is explicitly related to data recording. In general, a successful modeling of a wide variety of effects like inertia, finite resolution, amplification/damping that give rise to "non-linearities" in the recording, entails into partitioning of a BIS into windows of variable length. As a result, the "recording" constitutes a new stochastic sequence the members of which are the local averages with the length of the BIS rescaled by an arbitrary parameter n. But is it indeed another BIS?! Yes, it is. To prove it, let us have a closer look at the parameters that govern the value of a local average. It reads:

$$\overline{x}(\tilde{t}) = \frac{1}{(T_{i+1} - T_i)} \int_{T_i}^{T_{i+1}} x(t)dt \qquad (1.3)$$

where $T_i = T_{i+1} - T_i$ is the length of the $i-th$ window and $\tilde{t}$ is the rescaled time. Our first task is to show that new stochastic sequence remains bounded. Let r_{corr} be the correlation radius of the fluctuations. The largest possible value of

$\int_{T_i}^{T_{i+1}} x(t)dt$ cannot exceed $r_{corr} x_{thres}$ at $T_i > r_{corr}$.

Then, all the deviations of the local averages from the expectation value remain bounded within the following margins:

$$0 < \overline{x}(\tilde{t}) - \langle x \rangle \le \frac{r_{corr} x_{thres}}{T_{min}}. \qquad (1.4)$$

So indeed the new function is bounded if only T_{min} is non-zero.

Furthermore, the value of each member of the offspring BIS can be made equal to any *apriori* given one by an appropriate choice of the corresponding window length. Hence, the possibility for "designing" an offspring BIS is due to the variability of the window length. However there is another possibility for "designing". The local averaging maps an entire interval (window) onto a single point as to associate the value of the local averaging with this point. However, the mapping is not bijective since each point of the window is possible target. This is an immediate outcome of our supposition that all the scales and all the points uniformly contribute to the properties of the BIS. Thus, we are free to associate the averaged value with any point in the window. However, each particular choice constitutes a different offspring BIS. As a result the "digitizing" of a continuous signal (function) is mapped into discrete sequence the detail of which is not uniquely specified.

In summary, the "coarse-graining" can indeed modify every parameter that characterizes a given BIS, but leaves intact the property of the BIS that is to be bounded. I call this property transformational invariance. In more abstract terms it reads: the boundedness of a BIS is invariant under coarse-graining. Furthermore, the transformational invariance implies and justifies the transitivity of a set of BIS, i.e. each BIS can be reached by any other by the application of a sequence of appropriate coarse-grainings.

It should be stressed that the transformation of one BIS into another does not interfere with their self-similarity. Let me recall that self-similarity is a property that has nothing to do with the coarse-graining! Indeed, the self-similarity involves a simultaneous rescaling of the offspring function and the time by the same parameter in linear and homogeneous way, i.e. it acts on each member in the same *linear* manner. On the contrary, coarse-graining is essentially "non-linear" and its action is non-homogeneous.

Our amazing story about the coarse-graining is not yet finished. The first part was about how it transforms one "beast" into another. The next one will be about how it makes all the "beasts" into the same kind. In other words, we shall show that the BIS are scaling invariant.

Let us have closer look at the upper bound in Equation(1.4). Since both r_{corr} and x_{thres} have fixed values determined by the particularities of the BIS, the only parameter that governs the value of the local average is T_{min}. Evidently, when it monotonically increases, the term $\frac{r_{corr} x_{thres}}{T_{min}}$ becomes smaller and smaller. In turn, the margins within local averages deviates from the expectation value become more and more narrow. So, the first conclusion is that the expectation value does exist and it is finite. The second one is that local average is scale invariant. The next task is to prove this. According to the widespread interpretation, the scaling invariance is validated when the following relation holds:

$$\frac{1}{T}\int_0^{T/\delta} dt \int_t^{t+\delta}\left(x(\xi)-\langle x\rangle\right)d\xi = \varepsilon\left(T\right) = \varepsilon\left(\delta\right)\varepsilon\left(\frac{T}{\delta}\right).$$
(1.5)

Decoded into simple words, this reads: given a window of length T equipartitioned into subwindows of length δ, the local averaging can be done in two ways. The first one is by averaging

over the original BIS. The second one involves two steps. The first one is the averaging over the subwindows due to coarse-graining that results in an averaging over the subwindows and time rescaling by parameter δ so that a new BIS is produced. The second step is the averaging over the offspring BIS. The claim of Equation(1.5) is that the result should not depend on the choice of the parameter δ. Formally, it is achieved if and only if all the ε-functions are power functions of the same power and the sequence is scale-free, i.e. there is no scale (length) that has a specific contribution to the average.

But, the relation:

$$\varepsilon\left(\delta\right)\varepsilon\left(\frac{T}{\delta}\right) = \varepsilon\left(T\right)$$
(1.6)

seems unclear because the association of the ε-functions with the local average renders them ill-defined. Indeed, the local average varies from one window to another. Yet, the very meaning of the ε-functions as defined by the *l.h.s* of (1.5) is that they are to be associated with the local average! The only way out goes via the identification of the ε-functions with the upper limit of the local average determined by Equation(1.4). Indeed, since $\varepsilon\left(\delta\right)$ is local average, it cannot exceed $\frac{r'_{corr} x'_{tresh}}{\delta}$. By the same reasoning, $\varepsilon\left(\frac{T}{\delta}\right)$ cannot exceed $r''_{corr} x''_{tresh} \frac{\delta}{T}$ and $\varepsilon\left(T\right)$ cannot exceed $\frac{r_{corr} x_{tresh}}{T}$. Now, the substitution of the corresponding upper limits into (1.6) yields:

$$\frac{r'_{corr} x'_{tresh}}{\delta} r''_{corr} x''_{tresh} \frac{\delta}{T} = \frac{r_{corr} x_{tresh}}{T}$$
(1.7)

Evidently, the relation (1.7) confirms the scaling invariance of the original BIS, i.e. the independence from partitioning parameter δ. However, (1.7) says more than that: after the reduction of the scale parameters δ and T, it becomes an explicit expression of transformational invariance! Indeed, it relates the specific parameters r_{corr} and x_{tresh} of the original BIS and the specific parameters r''_{corr} and x''_{tresh} of the offspring BIS obtained by the coarse-graining.

So, the coarse-graining applied to a BIS entangles the transformational with the scaling invariance! Remarkable!

Let us now consider another aspect of the scaling invariance. We assert that (1.5) is a necessary condition for having the uniform convergence of the local average to the expectation value at $T \to \infty$. Intuitively, it seems that this is redundant because compliance with (1.4) is enough for this purpose. However, (1.4) is enough for the original BIS because its parameters r_{corr} and x_{tresh} are well-defined by our choice. But, how about its off-springs? Do they also have well-defined correlation size and thresholds of stability? The relation (1.7) gives an affirmative answer because it explicitly relates the parameters of the original and the offspring BIS. Since (1.7) allows a range of $\left(r_{corr}, x_{tresh} \right)$ combinations, it gives rise to an infinite number of well-defined BIS that can transform into each other. So, the scaling invariance validates the uniform convergence of the local average to the expectation value for the entire set of the BIS.

It is to be anticipated that the entanglement between the scaling and transformational invariance is what gives rise to those properties that are insensitive to the particularities of the BIS. Indeed, if one such property is established for a BIS subject to a scaling invariance, the relation (1.7) insures that this property is automatically shared by the entire set of its off-springs.

We now start the study of those properties of the BIS that are insensitive to the details of the fluctuations. The pivotal point is that the desired properties emerge from interplay of the boundedness and the scaling invariance but remain independent of the statistics of the fluctuation succession. As the scaling invariance is to play an important role we need first to investigate the properties of a BIS in a window of arbitrary but finite length T and then to take the limit $T \to \infty$.

Autocorrelation Function

We start with a very important characteristic of every irregular sequence, namely its autocorrelation function whose definition reads:

$$G\left(\eta, T\right) = \lim_{T \to \infty} \frac{1}{T} \int_{0}^{T-\eta} \left(X\left(t+\eta\right) - \langle x \rangle\right)\left(X\left(t\right) - \langle x \rangle\right) dt.$$

$$(1.8)$$

The autocorrelation function is a measure of the average correlation between any two points in the sequence separated by a time interval $\eta \leq T$. An intriguing interplay of the boundedness and the scaling invariance will be developed. On the one hand, the uniform contribution of all time scales restrains persistence extent without signaling out any specific one. On the other hand, the boundedness limitates the persistence because every deviation inevitably "turns back". Note that the uniform contribution of all time scales is very different from random contributions. The latter is characterized by the lack of any systematic correlations whilst the former is characterized by persistent but parameter-free correlations. But how far away is the persistence spread? And how the boundedness affects its extent?

The uniform contribution of all time scales gives rise to the following estimation of the autocorrelation function. An immediate outcome of (1.4) and (1.7) is that the local average in a window of length T is of the order of $\frac{1}{T}$. This means that the "distance" between the successive "zeroes" of the BIS, i.e. the points where the fluctuations cross the average, is bounded and is independent of T. This result applied to the autocorrelation function $G(\eta, T)$ yields the following estimation of $G(\eta, T)$:

$$H\left(\frac{\eta}{T}\right) \propto \sigma \lim_{T \to \infty} \frac{T - \eta}{T} = \sigma \lim_{T \to \infty} \left(1 - \frac{\eta}{T}\right)$$

(1.9)

where σ is the variance of the fluctuations. But how well does $H\left(\dfrac{\eta}{T}\right)$ approximate $G(\eta, T)$? The advantage is that $H\left(\dfrac{\eta}{T}\right)$ depends on η and T only through their ratio. This immediately makes the autocorrelation function independent from the particular choice of T. Note that $\dfrac{\eta}{T}$ is always confined to be in the interval $[0,1]$! In turn, the independence of the autocorrelation function $G(\eta, T)$ from T meets the requirement to be scale-free. However, the evasive point is that the permanent boundedness does not participate to $H\left(\dfrac{\eta}{T}\right)$?! This statement needs a clarification since so far we have been considering only *bounded* sequences!

The fact is that the limitation over the size of the distance between the zeroes is not unambiguously related to the permanent boundedness. Indeed, a bounded distance between the zeros can be sustained by the unbounded sequences as well - through making the rate of developing of the infinite fluctuations appropriately large. However, for the unbounded sequences, the only factor that governs the distance between the zeroes is the rate of the fluctuations development. On the other hand, the boundedness straightforwardly introduces an additional factor - it "enforces" the fluctuations to make a "U-turn" at the thresholds of stability regardless to their current rate. So, the "U-turns" increase the frequency of the zeroes. In turn, the

growth of that frequency results in an additional reduction of the long-range correlations. Note that the intensity of the long-range correlations is inversely proportional to the frequency of the zeroes.

To outline, the action of the boundedness on the autocorrelation function results in weakening of the long-range correlations. Note, however, that this weakening is not result of any physical process!

Hence, the additional reduction of the long-range correlations renders the autocorrelation function to decrease faster than $H\left(\dfrac{\eta}{T}\right)$. But how faster? In order to find out it we suggest that provided the limit $T \to \infty$ is taken, the shape of $G(\eta, T)$ reads:

$$G(x) = \sigma\left[1 - (x)^{\nu(x)}\right]$$

(1.10)

where $x = \dfrac{\eta}{T}$ and $x \in [0,1]$. The suggested shape of $G(x)$ meets the requirement that the autocorrelation function is parameter-free. Our next task is the specification of $\nu(x)$. To begin with, let us get use of $H(x)$. It is good approximation for both $G(x)$ and $\dfrac{dG(x)}{dx}$ at $x \approx 0$. So, by means of $G(0) = H(0)$ and $\left.\dfrac{dG(x)}{dx}\right|_{x=0} = \left.\dfrac{dH(x)}{dx}\right|_{x=0}$ its contribution to the specification of $\nu(x)$ reads $\nu(x) = 1 \pm px^n$. Further, the selection of p and n is made on the grounds of the requirement for uniform contribution of all time scales. The latter demands non-transversal approach of $G(x)$ at x going to 1. Then $\dfrac{dG(x)}{dx}$ goes to zero when x goes to 1. So, the non-transversality sets $\nu(x) = 1 - x^n$, but leaves n arbitrary. The latter is determined by the circumstance that the boundedness is not associated with any specific physical process. Indeed, note that while $\dfrac{dG(x)}{dx}$ monotonically decreases on x going to unity for every n, already $\left|\dfrac{d^2G(x)}{dx^2}\right|$ monotonically increases for every n apart from $n = 1$. Obviously, any increase of $\left|\dfrac{d^2G(x)}{dx^2}\right|$ interferes with our suggestion about the lack of any physical process associated with the course of the long-range correlations. So, our evaluations select $\nu(x) = 1 - x$ as the only exponent that fits all the requirements. Summarizing, the obtained exponent is the only one that ensures a non-transversal approach of $G(x)$ to zero so that all time scales perform uniformly and no special physical process is associated with the boundedness.

Summarizing the shape of autocorrelation function of a BIS reads:

$$G(x) = \sigma\left[1 - (x)^{1-x}\right] \qquad (1.10a)$$

where $x = \dfrac{\eta}{T}$ and $x \in [0,1]$

Note that the above derivation of the shape of the autocorrelation function of a BIS is explicitly grounded on the following two assumption alone: assumption about uniform contribution of all time scales and on the assumption that no physical process is initiated behind a "U-turn". It should be stressed that at the same time it does not require and does not involve any information about the statistics of a BIS. This makes the obtained property, i.e. the shape of the autocorrelation function, universal, i.e. insensitive to the nature of the BIS and to the particularities of the fluctuation succession. Thus, the time series that comes from systems as diverse as quasar pulsations, DNA sequences, financial time series etc. share the same autocorrelation function.

POWER SPECTRUM

Another important characteristic of every irregular sequence is its power spectrum $S(f)$. Though it is the Fourier transform of the autocorrelation function $G(\eta, T)$, it cannot be evaluated by the straightforward application of the Fourier transformation. To elucidate the difficulties, let us have a look at the definition:

$$S(f) = \int_0^1 G(x)\exp(-ifx)dx =$$
$$\int_0^1 \left(1 - x^{1-x}\right)\exp\left(-ifx\right)dx \qquad (1.11)$$

The usual trick to estimate such integrals is to rescale the variable x to $x' = fx$. The purpose is to make the integral free from f and to collect the dependence on f into term that multiplies the integral. However, now this trick the does not work because the non-constant exponent $\nu(x) = 1 - x$ prevents the "elimination" of the f-dependence from the integral. Indeed:

$$S(f) = \frac{1}{f}\int_0^f \left[1 - \left(\frac{x'}{f}\right)^{1-\frac{x'}{f}}\right]\exp\left(-ix'\right)dx' \qquad (1.12)$$

Intuitively, the problem seems only technical. However, it has a fundamental aspect as well because the Fourier transformation is an essentially non-linear operation. So it is likely to expect local "amplifications", "stretching" etc. Yet, the very definition of the operation makes these non-linearities to appear only when certain time scale has specific contribution. Let us just remember that the Fourier spectrum of every periodic function is discrete and its components are concentrated around its periods and their harmonics. Indeed, any long-range correlation between two time scales appears as single line whose amplitude is proportional to their correlation. Recalling that on the one hand, the uniform contribution of all time scales restrains persistence extent without signaling out any specific one and on the other hand, the boundedness limitates the persistence because every deviation inevitably "turns back", the question becomes how the interplay of the long-range correlations introduced by the boundedness and the scaling invariance of the time scales shapes the power spectrum. Our first expectation is that the power spectrum must be a monotonic function so that not to signal out any specific component. Indeed, a closer look at (1.12) indicates that the power spectrum is a strictly decreasing power function. But is the exponent again non-constant? And if so, is their some property of that exponent that remains invariant under the Fourier transformation? Further, is the exponent again insensitive to the statistics of the BIS?

Equation (1.12) indicates that $S(f)$ is a decreasing power function with non-constant exponent:

$$S(f) \propto \frac{1}{f^{\alpha(f)}} \qquad (1.13)$$

where the shape and the properties of $\alpha(f)$ are to be worked out. The accomplishment of this task is made on the basis of the permanent boundedness of $X(t)$ by the use of the Wiener-Khinchin theorem. It relates the power spectrum $S(f)$ and $y_T(f)$, i.e. the Fourier transform of $(X(t) - \langle x \rangle)$ over a window of length T. More precisely, the relation reads:

$$S(f) = \lim_{T \to \infty} \frac{1}{T} \left\langle \left| y_T(f) \right|^2 \right\rangle \qquad (1.14)$$

Actually, since the power spectrum is discrete for any finite T, (1.14) states that $\frac{1}{T} \left| y_T(f) \right|^2$ uniformly fits the shape of $S(f)$ as T approaches infinity. So, $X_T(t)$ is majorized by:

$$I_T(t) = \sqrt{T} \int_{1/T}^{\infty} \frac{\cos ft}{f^{\alpha(f)/2}} df + \sqrt{T} \int_{1/T}^{\infty} \frac{\sin ft}{f^{\alpha(f)/2}} df \qquad (1.15)$$

where the Fourier coefficients are constructed on the basis of (1.13) and (1.14). In order for $I_T(t)$ to serve as an estimate of $X_T(t)$ it should be finite for every t and T. It is worth noting that the latter is mathematical expression of the view that boundedness is property that operates permanently in a time interval of unspecified length. For this purpose it is enough to find out at what values of $\alpha(f)$

$$I_T(0) = \max \left| I_T(t) \right| = \sqrt{T} \int_{1/T}^{\infty} \frac{1}{f^{\alpha(f)/2}} df \qquad (1.16)$$

is finite and its value does not depend on T.

The integration of a power function of non-constant exponent is highly contraintuitive and involves a non-trivial step. We account for the details in the Appendix at the end of this chapter because their presentation is rather lengthy and stays a little apart from the mean stream of the section. Though, we urgently recommend its studying.

Further, we utilize the obtained result, namely:

$$\int_{a}^{b} \frac{df}{f^{\alpha(f)}} = \frac{b^{-\alpha(b)+1}}{1 - \alpha(b)} - \frac{a^{-\alpha(a)+1}}{1 - \alpha(a)} \qquad (1.17)$$

Actually, the application of (1.16) to the *r.h.s.* of (1.17) selects that $\alpha(f)$ which renders $I_T(0)$ to be bounded in margins that are independent of T. The formal expression of this property is through:

$$I_T(0) = \lim_{T\to\infty} \sqrt{T} \int_{1/T}^{T} \frac{df}{f^{\alpha(f)/2}} =$$

$$\lim_{T\to\infty} \sqrt{T} \left. \frac{f^{1-\frac{\alpha(f)}{2}}}{1-\frac{\alpha(f)}{2}} \right|_{1/T}^{T} =$$

$$= \lim_{T\to\infty} \frac{T^{\frac{3-\alpha(T)}{2}}}{1-\frac{\alpha(T)}{2}} - \lim_{T\to\infty} \left(\frac{1}{T}\right)^{\frac{1-\alpha\left[\frac{1}{T}\right]}{2}} \frac{1}{1-\frac{\alpha\left(\frac{1}{T}\right)}{2}} = T^0$$

$$(1.18)$$

Evidently, the second line of (1.18) holds when $\alpha(f)$ is a continuous strictly increasing function between the following limits:

$$\alpha\left(\frac{1}{T}\right) = 1 \text{ and } \alpha(\infty) = p \qquad (1.19)$$

where p is arbitrary but $p \geq 3$.

But we can go further in the specification of $\alpha(f)$! The scaling invariance applied to the power spectrum implies that neither of its frequencies has specific contribution. This is another way to say that all time scales contribute uniformly in the power spectrum. Therefore, neither $\alpha(f)$ nor any of its derivatives should select any specific time point. Simple calculations show that it is possible if and only if $\alpha(f)$ is a linear function. Any non-linearity in $\alpha(f)$ makes its derivatives to change their sign at certain frequency. However, the change of the sign makes the corresponding time scale specific and requires additional physical process that "makes" the sign to change.

Above we found out that both $\nu(x)$ and $\alpha(f)$ are strictly monotonic linear functions of their arguments. In both cases the strict monotonicity along with the linearity of the exponents comes out from the uniform contribution of all time scales. Therefore the linearity of the exponents is property invariant under the Fourier transformation. Indeed, non-trivial and unexpected!

Note that the derivation of the non-constant exponent $\alpha(f)$ does not involve any reference to the statistics and the length of the BIS. This allows concluding that the shape of the power spectrum of the BIS is also universal, i.e. it is insensitive to the details of the fluctuations succession and is invariant under the length of the time series. These properties render it as the first of our time-series invariants. It is worth noting that this property is indeed exclusive for BIS`s since the shape of the power spectrum of an *unbounded* time series explicitly depends on the statistics of its members and its length!

So, the power spectrum of each BIS has the following properties:

1. The spectrum has an infrared cut-off at the frequency $f_{min} = \frac{1}{T}$ where T is the length of the time series;

2. $\alpha(f)$ is linear function so that $\alpha(f) = 1 + \kappa(f - f_{min})$ where κ is a specific to the system parameter that is strongly related to the properties of the "U-turns".

Actually, κ is proportional to the frequency of "U-turns". The major contribution to that frequency comes from the probability that the current fluctuation reaches the thresholds of stability. According to the Lindeberg theorem (Feller, 1970) every BIS of infinite length has finite expectation value and finite variance. As a result, the fluctuation statistics approaches normal distribution truncated by the thresholds of stability. Thus:

$$\kappa \propto \exp\left(-\frac{X_{thres}^2}{\sigma^2}\right) \qquad (1.20)$$

It seems that the persistent shape of the power spectrum give an easy way to determine the value of the thresholds of stability. The way goes through the extraction of κ from the power spectrum. However, nobody has ever proceeded this way! The answer is that for a lot of practical cases κ is extremely small. To illustrate the smallness of κ, let us have a look at Table 1:

Evidently, the smallness of some κ goes beyond the most elaborate ways for its extraction from the power spectra. Then what we have in hand from the power spectra that come from the experimental records. Do they support our considerations? Above we have established that the power spectrum of every BIS fits the same shape regardless to the details of the fluctuation succession and the length of the window. Does the physical world support our considerations?!

Indeed, the power spectra of all fluctuating systems persistently fit the same shape established to be $1/f$ in large frequency interval. The persistency of the shape even named the phenomenon - $1/f$ noise. This happens to be one of the most ubiquitous and widely studied phenomena in the world. Though it has been systematically studied for more that a century, two of its major properties remain mystery. The first one is the insensitivity of the power spectrum shape to the details of the fluctuations statistics. One of the most striking examples is that the power spectra of both traffic noise and a Beethoven symphony fit the same shape! The second one is the insensitivity of the shape to the length of the time series. It means that if the power spectrum of some time series of length T fits $1/f$ shape, the power spectrum of the same time series but of different length T_1 fits the same shape. Moreover, T_1 can be made dozens of orders greater than T.

Evidently, these enigmas are successfully revealed by the considerations in the present section. However, the problem whether the exponent α is constant or function is crucial for the conjecture of boundedness! Indeed, let us come back to (1.17). Simple calculations show that if α is constant, neither its value can provide permanent boundedness. Indeed, an ultraviolet divergence emerges for $\alpha \leq 1$, while for $\alpha > 1$ there is an infrared one. So, only an exponent that changes with the frequency can provide permanent boundedness! Besides, if the fluctuations were un-

Table 1.

$\dfrac{X_{thres}}{\sigma}$	$\left(\dfrac{X_{thres}}{\sigma}\right)^2$	κ
2,25	5	0.006
3	10	10^{-5}
4	16	10^{-7}
4.5	20	10^{-9}
7	50	10^{-22}

bounded, they would go beyond the thresholds of stability that would result in collapse of the system! On the other hand, that $1/f$ noise has been spanned over several dozens of orders. Hence, it is rather to be related to the stability of the system. This is an important point in our score! But the major one, the determination of κ is still to come! Now we shall present an elegant way for its determination.

VARIANCE

Setting of κ

A crucial test for all our considerations is the evaluation of the power spectrum amplitude. Indeed, we can make it into two ways - once through its definition:

$$\sigma = \lim_{T \to \infty} \frac{1}{T} \int_0^T \left| X(t) \right|^2 dt \qquad (1.21a)$$

and on the grounds of the spectral density (power spectrum):

$$\sigma_s = \lim_{T \to \infty} \frac{1}{T} c \int_{1/T}^{\infty} \frac{df}{f^{\alpha(f)}} \qquad (1.21b)$$

where c is the amplitude of the power spectrum that is to be determined. The permanent boundedness of $X(t)$ automatically provides the existence of variance calculated by means of (1.21a). However, σ_s is problematic since:

$$J = \lim_{T \to \infty} \frac{1}{T} \int_{1/T}^{\infty} \frac{df}{f^{\alpha(f)}} \qquad (1.22)$$

seems to have singularity in its lower limit, namely:

$$J = \lim_{T \to \infty} \frac{1}{T} \frac{\left(\frac{1}{T} \right)^{1 - \alpha\left(\frac{1}{T} \right)}}{1 - \alpha\left(\frac{1}{T} \right)} = \lim_{T \to \infty} \frac{1}{T} \frac{T^0}{0} \qquad (1.23)$$

To reveal the singularity we should determine the rate by which $\alpha(f)$ goes to 1. For this purpose let us recall that the power spectrum of every BIS of arbitrary but finite length is discrete. So, the distance between any of its two components cannot be made smaller than $\frac{1}{T}$. So, the rate by which $\alpha(f)$ approaches unity is proportional to $\frac{1}{T}$. Let us now evaluate J_1:

$$J_1 = \lim_{T \to \infty} \frac{1}{T} \int_{2/T}^{\infty} \frac{df}{f^{\alpha(f)}} \qquad (1.24)$$

The purpose is twofold: on the one hand, when T goes to infinity, J_1 converges to J. On the other hand, the "shift" of the lower limit in J_1 allows involving the rate by which $\alpha(f)$ goes to 1. Indeed:

$$J_1 = \lim_{T \to \infty} \frac{1}{T} \frac{\left(\frac{1}{T} \right)^{1 - \alpha\left(\frac{2}{T} \right)}}{1 - \alpha\left(\frac{2}{T} \right)} = \lim_{T \to \infty} \frac{1}{T} \frac{1}{\kappa \frac{1}{T}} \left(\frac{1}{T} \right)^{\frac{k}{T}} =$$

$$\lim_{T \to \infty} \frac{1}{\kappa} \frac{1}{T} \left(\frac{1}{T} \right)^{-\frac{\kappa}{T}} \qquad (1.25)$$

How annoying! The removal of one singularity brings about another one!? What the behavior of x^{-x} is at x going to zero? To find out, we use the exponential presentation, namely:

$$x^{\pm x} = \exp\left(\pm x \ln x\right) \tag{1.26}$$

Further, since x goes faster to zero than $\ln x$ diverges:

$$\lim_{x \to 0} x^{\pm x} = \exp\left(\pm \lim_{x \to 0} x \ln x\right) = 1 \tag{1.27}$$

Finally, the variance calculated on the grounds of the spectral density reads:

$$\sigma_s = \frac{c}{\kappa} \tag{1.28}$$

Since the variance can be determined independently by means of (1.21a), (1.28) sets the value of c, i.e. the amplitude of the power spectrum. As a result we have established all the characteristics of the power spectrum and its final specification reads:

$$S\left(f\right) = \sigma \kappa \frac{1}{f^{\alpha(f)}} \tag{1.29}$$

where $\alpha\left(f\right) = 1 + \kappa\left(f - \frac{1}{T}\right)$ and $f \geq \frac{1}{T}$.

Now we can utilize the last property, namely the existence of infrared cutoff of the power spectrum. A closer look at (1.29) shows that the amplitude of the cutoff component is $A_c = \sigma \kappa T$. This gives any easy way to extract κ: it is proportional to the slope of the plot of A_c on the length of the time series T. The linearity of that plot makes the task particularly easy.

The proportionality of the power spectrum amplitude on κ reconciles another mystery of the $1/f$ noise. It has been established that $1/f$ spectrum is well discerned only for extremely long time series! A brief look at the amplitude of the cutoff component $A_c \approx \kappa T$ tells that it becomes of the order of unity at $T \propto \frac{1}{\kappa}$. And the extreme smallness of κ explains why the mysterious and ubiquitous $1/f$ noise emerges only if one is patient enough to record very long time series! And this is another point in our score! To make the things more clear let us suppose that the amplitude of the power spectrum does not comprise κ and is of the order of unity. Then $A_c \propto T$. Further, suppose that the amplitude of a specific for the system process is also unity. Consider now a record that comprises 10^3 points. Keeping in mind that the variance is unity, the amplitude of the noise band would be thousand times greater than that of any specific to the system process!? And how about a record of billion points? Thus, thanks to the extreme smallness of κ, the "noise" band in the power spectra can be separated from the contributions that comes from the specific for a system processes. However, the noise band hides extremely valuable specific information - that about the thresholds of stability. The practical importance of that information is enormous: there is no need to wait for a "bump" into the thresholds of stability to find out whether a collapse is to be expected.

BOUNDEDNESS OF THE RATES: EMBEDDING DIMENSION

So far we have established that the power spectrum and the autocorrelation function of the BIS exhibit remarkable universality – under the mild constraint of permanent boundedness alone we have proved that their shape and characteristics are free from any specification of the fluctuation dynamics. In turn, our mathematical proof gains strong support by the enormous diversity of systems who share the same ubiquitous $1/f$ shape of their power spectra; among these systems are earthquakes, traffic noise, Beethoven symphonies, DNA sequences to mention a few. Though this ubiquitous insensitivity has been rigorously proven mathematically and supported by the enormous diversity of observations, our intuition remains unsatisfied and we are strongly tempted to suggest that the power spectrum and the auto-correlation function are too "coarsen" character-istics to take into account the particularities of the fluctuations dynamics. Indeed, the dynamics of the fluctuations comes out from the specific for every system processes associated with their development. Yet, now we will show that the fluctuation dynamics is also subject to certain universality.

The usual way to reveal the dynamics of the correlations in a stochastic sequence is to study the properties of its phase space. The procedure involves dividing of the phase space volume into small size cells and counting the points at which the trajectory intersects each of them. It is certain that this elaborate operation helps much in the ex-amination of the correlation dynamics. Alongside, it is to be expected that the obtained information is highly specific. However, the general aim of this section is to study the universal properties of the BIS. That is why now we focus our attention on the question how the boundedness is incorporated in the dynamics of the correlations.

We start with substantiating the general con-straint that the boundedness imposes on the fluc-tuation dynamics. The notion about boundedness is associated with the conjecture that the each object/subject comprises finite amount of energy and matter. The boundedness of the fluctuation size is one of its manifestations. However, it is not enough to keep a system stable. Likewise it is necessary that the amount of energy/matter exchanged with the environment is also permanently limited. This requirement imposes permanent boundedness of the rate of the fluctuations development. Further I call the boundedness of the fluctuation rate dy-namical boundedness while the boundedness of the fluctuation size is named static boundedness.

Our expectation is that the interplay of the dynamical and static boundedness renders every

BIS confined in a finite phase space volume of specific dimension. It is easy to understand how the limitation over the volume of the attractor appears. Indeed, since every BIS is comprised by bounded variations around the expectation value, the attractor is a compact set that has a fixed point matched by the expectation value. The upper bound in each direction is set on the thresholds of stability. So, the static boundedness indeed provides finite volume of the attractor but it is not enough to prompt its dimension.

The specification of the phase space dimension looks hard and confusing. The usual practice is to make guess on the grounds of the temporal be-havior of a single or a few variables. But, does the guess involve specific arguments in each particular case or there is a general rule?! My next task is to show how the interplay of the static and dynami-cal boundedness provides finite dimension of the attractor whose value is specific to the dynamics of the BIS but always finite.

To begin with, let us mention that the dy-namical boundedness makes every fluctuation extended in the time course. More precisely, it means that every fluctuation is approximated by a trajectory that starts at the expectation value at a given moment, pass trough maximum and turns back to it for the first time after certain time interval called hereafter duration. Further, the duration of that interval is related to the size of the corresponding fluctuation so that the rate of development is permanently bounded. Hence, both the duration and the size of all fluctuations are finite and limited by the size and the duration of the largest one.

Now we are ready to start the study of the phase space properties. Our first task is to show that its dimension is always finite. Given a one-dimensional rescaled BIS $X(t) = \dfrac{\tilde{X}(t) - \langle \tilde{X} \rangle}{\sqrt{\sigma}}$ where $\tilde{X}(t)$ is the original BIS, $\langle \tilde{X} \rangle$ is its expectation value and σ is the variance. By means of this operation the only specific to the BIS parameter is its threshold of stability. Let us now apply the widely used in the time series analysis procedure of time delay embedding. The purpose is to make the comparison to the well established results straightforward. The time-delay embedding implies that the vectors $\vec{R}_n(t) = \left(X(t), \ldots X(t - i\tau), \ldots X(t - n\tau) \right)$ are embedded in a n dimensional Euclidean space. $X(t)$ are the successive points that come from a given time series; τ is a small delay. Further, the phase space is divided into small size cells and the vectors whose ends are inside each cell are counted. The population in lg-lg scale is plotted against the cell size. Applied to a BIS, the time delay embedding is a noting more than a particular way of coarse-graining. So, the ends of the vectors $\vec{R}_n(t)$ also construct a BIS. Then we are able to associate the value $\bar{X}(t - i\tau)$ of the coarse-grained BIS with the $i - th$ axis, where $i \in [1, n]$.

Note that the association of the delay $i\tau$ with the topological dimension is equivalent to the parameterization of the size of the fluctuation through the phase space angle. But the size of a fluctuation is already parameterized trough its relation with its duration. The evident parity of both parameterizations selects a topological dimension so that a fluctuation forms a closed continuous curve by a single revolt. The topological dimension that renders the largest fluctuation to appear as a loop is called embedding dimension. The hallmark of this dimension is that all the smaller fluctuations are closed continuous loops so that each loop has its own embedding dimension smaller than the embedding dimension of the entire attractor. And the boundedness of both the size and duration makes the embedding dimension always finite.

To make the above considerations more evident let us point out that when the topological dimension n is smaller than the embedding dimension, the larger fluctuations appear as complicate heli-

ces not as closed curves. Depending on the ratio between the fluctuation and the topological revolt angle, the helix is first unwinding then turns to winding. The unwinding happens whenever the current revolts are not enough for the fluctuation to reach its full size. The winding is associated with the development of the fluctuation from its full size to zero.

But how to illustrate definitively the established structure of the attractor?! Can we exhibit its property to stretch and fold the distance between successive points of the BIS? Indeed, when the helix unwinds, the distance stretches while its winding results in folding. That is why it is to be expected that the stretching and folding strongly affects the density of the attractor. But how? We start the task by establishing of the factors that govern the "stretching and folding" mechanism. The first one is the topological dimension n. It selects a size of the fluctuation l_n so that its embedding dimension coincides with n. So, the fluctuation whose size is l_n forms a loop. Thus l_n acts as the demarcation where stretching turns to folding. Put it in geometrical terms, the topological dimension drives the "angle" of the helix. The factor that drives the step of the helix is the finite rate of the fluctuations development. Indeed, though the helicoidal behavior is purely geometrical effect, it should integrate the limited rate of the fluctuations development. The incorporation of the limited rate into the helicoidal behavior is through the following non-linear relation of power type between the angle u and the step l:

$$l = l_n u^{\gamma_n(u)} \qquad (1.30)$$

where u is rescaled so that $u \in [0, 1]$, $\gamma_n(u)$ are chosen so that to fit two conditions, the first of which is that $l = l_n$ at $u = 1$. Actually, this condition means that the helix turns to loop of size l_n by a single revolt. The second one is that it should provide specific for the system dynamical rate. Further, it should provide the same dynamical rate at each topological dimension. Simple geometrical considerations show that it is achieved by setting $\gamma(u)$ dependent on the topological dimension n. It is worth mentioning also that the meeting of the above conditions makes $\gamma(u)$ to increase on the increase of n. Note that the power type relation between the step and the angle of the helix is parameter-free! In other words, it does not select any size that has specific properties. Thus, neither scale has specific contribution to the stretching and folding mechanism!

Now we come to the matter about the density of the points that are subject to stretching and folding. Evidently, the major factor that sets their participation is the correlation between the points. We have already found out that the average probability that two points separated by a time interval η have the same value is the autocorrelation function $G(x)$. Since both points have the same value, they can be considered as belonging to the same effective fluctuation whose duration is $x = \dfrac{\eta}{T}$. The size of the fluctuation $\tilde{l}$ is limited by l_n and the relation between x and $\tilde{l}$ is set on the dynamical boundedness. Further, the requirement that both the size and the rate of development of a fluctuation remain bounded without involving any specific process (scale) renders the power type relation between $\tilde{l}$ and x:

$$\tilde{l} = l_n x^{\tilde{\gamma}(x)} \qquad (1.31)$$

The apparent similarity between (1.30) and (1.31) is actually equivalence! Indeed, since both x and u are confined in the same range $[0,1]$, the same dynamical rate imposes $\gamma(u) = \tilde{\gamma}(x)$ for every $x = u$. The equivalence of (1.30) and (1.31) proves explicitly our heuristic arguments about the parity between the parameterization of

a fluctuation through its duration and through the phase space angle.

Therefore the probability that two points separated by distance l are subject to stretching and folding is $G(u)$. On the other hand, $(1 - G(u))$ is the probability that the distance remains l. Therefore, the probability that the distance between a pair of point is less than l reads:

$$C(x) = 1 - \int_{0}^{u} G(x)\,dx = 1 - P(u) \qquad (1.32)$$

where

$$P(u) = \int_{0}^{u} G(x)\,dx = \int_{0}^{u} \left(1 - x^{1-x}\right)dx = x - \frac{x^{2-x}}{2-x} \qquad (1.33)$$

The plot of the function $P(u)$ given in Figure 1 reveals peculiar behavior: it goes through maximum! Does it mean selection of a specific scale?! To uncover the mystery, let us start step by step: the increasing part of the plot is obvious - on the increase of u the correlation $P(u)$ increases until current size of the unwinding helix reaches its "loop" size l_n. Further the helix turns to winding. However, the winding shrinks the distance and so brings about an effective decrease of the correlations that participate stretching. Thus, the plot of $P(u)$ strongly supports the idea about the boundedness: the stretching is limited by certain size where it turns to folding. Note, that while expressed through u, the boundary between the stretching and folding is universal and independent of the topological dimension, its size l_n is specific to the attractor and depends on the current topological dimension n as well.

It is to be expected that the fine structure of the attractor is highly non-trivial and very different from the coarse-grained one. This statement is very well illustrated by the population structure.

Figure 1. The probability $P(u)$ for a pair to participate the stretching and folding

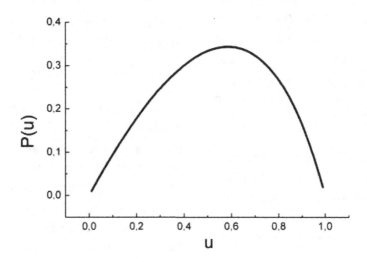

Indeed, let us divide the phase space into cells of size l and count all points inside each of them. The measure of the corresponding number is the plot of $C(l)$ vs. the size l where l is related to u trough (1.30). To reveal better the difference between fine and coarse-grained partitioning, the plot is presented in lg-lg scale. Note that since

$C(l) \equiv C(u)$ where $C(u)$ is set on (1.32), the plot is actually parameterized by u trough the relation (1.30) (Figure 2).

curve $1: l = 10.3u^{3.2}$ *; curve* $2: l = 11u^{4.5}$ *; curve* $3: l = 12u^{6.6}$.

Figure 2. The plot of the phase space population on the degree of coarse-graining

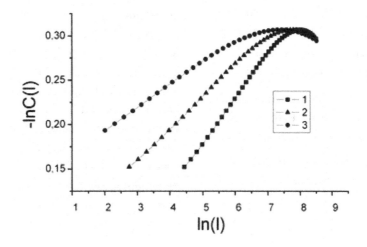

The values of l_n and γ_n in Figure 2 are set so that to correspond to different topological dimensions. Though we do not specify the actual topological dimension n, it is easy to find out that both l_n and γ_n grow on n increasing. Note that the growth is not arbitrary but l_n and γ_n are related so that to leave the dynamical rate intact. The plot exhibits the following characteristic properties of the population behavior:

1. At every topological dimension the plot manifests nearly linear part. Accordingly the population behaves as l^{-D_n} where D_n decreases on n increasing.
2. At every topological dimension the population reaches saturation;
3. The saturation value is insensitive to the value of the topological dimension.

The linear part of the plot is apparently associated with the prevailing role of the stretching. Further, since the folding shrinks the distance between the points, it effectively weakens the correlations necessary for stretching. In other words it acts towards stochastisation of the points. Thus the population tends to remain constant and insensitive to any further growth of the cell size. Furthermore, remember that $C(u)$ is independent of the topological dimension. This immediately renders the saturation value insensitive to the topological dimension as well. However, the plot of $C(u)$ vs. cell size l is good approximation for the population only up to the saturation. It should be remembered that $C(l)$ is the measure for the correlations that give rise to the stretching. However, since the folding contributes to the stochastization, $C(l)$ ceases to be the measure of the population for the cell sizes larger than the cell size at which the saturation emerges.

Let us now estimate the value of the cell size at which the saturation occurs. Let me recall that $C(l) \equiv C(u)$. An immediate look at $C(u)$ from (1.32) shows that the saturation is achieved at u very close to 0.6. Simple calculations show that

$$r(u) = \frac{l(u = 0.6)}{l_n} = 0.03$$ on the curve 3. So,

the population changes its behavior at cell size that is only 3% of l_n! Hence the fine structure of the attractor becomes evident only at very fine partition of the phase space. Yet, note that the value of $r(u)$ strongly depends on the rate of development of the fluctuations.

So far so good. But do the time series that come for the real events match such behavior? Yes, they are. There is plenty of time series coming from a broad spectrum of systems whose population is studied by means of the time-delay embedding and exhibits exactly the same behavior as that depicted in Figure 2. Moreover, this behavior is the one recently taken as decisive for the deterministic chaos!

Though the chaos has been intensively studied, a rigorous and non-ambiguous definition is still missing. Our ambition is to illustrate that alone the static and dynamical boundedness give rise to the major chaotic properties established so far. The greatest advantage is that the physical conjecture about the boundedness is available for a wide variety of systems. This corresponds to the enormous ubiquity of the chaotic behavior. It is typical for phenomena of very different nature including engineered, natural and social ones. So, if there is a general idea behind the chaos it must give rise to all its aspects established so far. Our first task in this direction is to prove that the boundedness alone indeed brings about its major characteristics: autocorrelation function, power spectrum and embedding dimension from the viewpoint of the boundedness.

K-ENTROPY

The task of the present section is to demonstrate that the boundedness not only brings about another major characteristic of the deterministic chaos, K-entropy, but the latter appears as a time-series invariant.

Historically the deterministic chaos has been observed for the first time by Lorentz at numerical simulation of a simple system of 3 deterministic non-linear ordinary differential equations (ODE) that fits the Lipschitz conditions. He has established that at certain parameters its solution behaves in highly non-trivial way: it widely varies and exhibits strong sensitivity to initial conditions. This sensitivity implies that every infinitesimal deviation from the initial conditions diverges to finite size in finite time interval. Some authors choose this property as definitive for the chaos. Still, it lacks enough universality since it is straightforwardly related only to the deterministic dynamical systems. Moreover, the sensitivity towards the initial conditions suffers a great disadvantage - though nowadays it has clear explanation, the irregularities of the solution remain mystery. One of the greatest achievements of the concept of boundedness is its ability to explicate that the chaotic properties appear as a result of mere boundedness .

Let us start with the idea of the homoclinic orbits. It implies that the phase space is a dense set of orbits that has a fixed point and whose period is infinite. In turn, the homoclinic orbits give rise to a continuous power spectrum. Sounds familiar? Indeed, let us remember that the phase space of a BIS is a set of "loops" each of which is fixed to the origin. Remember that since the fixed point is the expectation value of the BIS, each "loop" is fixed to it. Further, the size of the loops continuously fills the range from zero to the thresholds of stability. The period of the loops is infinite - though each loop is repeated in finite time intervals, its *period* is infinite. Let us recall that the notion of a period implies a regular repetition in a given period while the repetitions of the loops succeed in finite time intervals but following in an irregular manner. Thus, the association of the loops with the homoclinic orbits is straightforward. Moreover, the power spectrum of a BIS is also continuous band. Note, that the infinity of the "loops" periods substantiates the property of the power spectrum to be a continuous band! Yet, the infinity of all loop periods cannot set the monotonic decrease of the power spectrum. But the boundedness can! It specifies not only the monotonic decrease but settles the precise shape as well!

Another very important aspect of the chaoticity was considered in the previous section. It is believed that the phase space of each chaotic system has specific but finite dimension. Further, this dimension can be extracted from the specific behavior of the phase space population depicted in Figure 2. The missing point, however, is the interrelations among those three aspects of chaoticity. The lack of such interrelations gives rise to the following puzzle – one the one hand, the infinite periods of the homoclinic orbits render an infinite density of the attractor. Indeed, the folding of an arbitrary long time series in the finite volume of its attractor makes the density infinite. On the other hand, the behavior of the population depicted in Figure 2 implies infinite repetitions of the trajectory so that the number of the discernible points remains permanently finite and insensitive to the length of the time series. The major advantage of the idea about the static and dynamical boundedness is that it manages not only to unify the idea of the homoclinic orbits, continuous power spectrum and the embedding dimension but to reveal the above discrepancy. Let me first recall that though the "loops" have infinite periods, they repeat in finite but irregular time intervals. Further, since the distances between the zeroes of a BIS are finite and independent of the BIS length, the frequency of the repetitions is insensitive to the length of the BIS. Evidently, this consideration is enough to conclude that the population of the discernible points remains finite and insensitive to the length of the BIS.

Another very important and decisive characteristic of the chaotic behavior, not yet considered by us, is the so called Kolmogorov entropy or K-entropy for short. By definition it reads:

$$K = -\lim_{\tau \to 0} \lim_{l \to 0} \lim_{N \to 0} \frac{1}{N\tau} \sum_{i_0 \cdots i_N} P_{i_0 \cdots i_N} \ln P_{i_0 \cdots i_N}$$

(1.34)

where $P_{i_0 \cdots i_N}$ is the joint probability that the $\vec{R}(t=0)$ is in the i_0 cell, $\vec{R}(t=\tau)$ in the i_1 cell,..., $\vec{R}(t+n\tau)$ - in the i_N cell. The sum is taken over all possible partitionings of the phase space. This makes K measure of the average information necessary for precise setting of the motion in the phase space. The value of K-entropy is definitive for the chaos since it is zero for the deterministic motion infinite for the stochastic one and finite for the deterministic chaos. So, what is its value for the BIS?! Let us consider the function:

$$K(u) = -\left(1 - P(u)\right) \ln\left(1 - P(u)\right)$$

(1.35)

Since $\left(1 - P(u)\right)$ is the probability that any two points separated by time interval u are not stretched, $K(u)$ is the information necessary to fix that distance between the initial and the finial point to u. Note that this probability does not depend on the details of the trajectory. Since it is impossible to predict better the motion on the attractor, $K(u)$ covers the meaning of K-entropy. But is it finite?! Evidently, it is. To make it apparent, let us have a look on its plot demonstrated in Figure 3.

$K(u)$ has two very important properties. The first one is that its behavior is straightforwardly related to the stretching and folding. The explication of that relation becomes evident by setting $K(u)$ through $\left(1 - P(u)\right)$, where $P(u)$ is the probability for participation to the stretching and folding. Then the plot of $K(u)$ vs. u has the following meaning: the increase of the probability for stretching at small u requires more information for setting the distance between the initial and the final point of any trajectory. Further, the stretching reaches its maximum and turns to folding which, however, shrinks the distance between the initial and the final point of the trajectory. Thus the folding contributes to better precision of the position of the motion. In turn, better precision needs less information for specifying the characteristics of the trajectory. As a result, the K-entropy turns to decrease.

The second property is that we do *not* relate $K(u)$ to the so called Lyapunov exponent. Why? Let us first explain what Lyapunov exponent is. It is the measure of the stretching and folding presented in an exponential form. However, it should be stressed that this presentation is not parameter-free while the stretching and folding associated with the dynamical and static boundedness is presented by the parameter-free power form (1.31) and (1.32). Let us outline that the power type form is the only possible one that meets the interplay of the requirements about the lack of a specific time scale and the dynamical boundedness. On the contrary, though the exponential stretching does not select specific time scale, it is not able to meet the dynamical boundedness. So, the correct definition of the Laypunov exponent must take into account the dynamical boundedness. We postpone the consideration until Chapter 5 since then we will be able to as-

Figure 3. K-entropy for a BIS

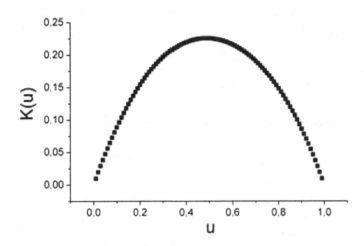

sociate explicitly the Lyapunov exponent with the automatic execution of the U-turns.

Yet, the property that makes K-entropy a time-series invariant is the universality of its shape to the details of the time series. Moreover, its non-zero value for all arguments actually puts a universal bound over the possibility for prediction of the exact future behavior of any complex system. On the other hand, the ban over precise predictability completely corresponds with the assumption about the release from any global pre-determination of the impact and the response. The highest non-triviality of the matter about the predictability is that the only predictable variables are those associated with the homeostasis while the local "fluctuations" are left non-specified, yet being subject of the mild assumption of boundedness.

Outlining, the universality of K-entropy has two-fold comprehending: (i) on the one hand, it put a universal bound over the finest precision of predicting detailed future behavior of a BIS; (ii) at the same time it provides universal finite rate of predictability for existence of a homeostatic pattern.

The matter how to predict any change in homeostasis by means of the time series invariants along with other important properties of the time series is considered in the next chapter.

CONCLUSION

So far we have established that the static and the dynamical boundedness (boundedness of rates) indeed make every BIS to display time-series invariants. We have proven that there are 3 characteristics of a BIS which are insensitive to the statistics of its members. They are: the shape of the autocorrelation function, the shape of the power spectrum and the shape of the K-entropy; along with them comes the universality of the population of an attractor. It is worth noting that the existence of the time-series invariants is an exclusive property of boundedness which is not shared by any *unbounded* counterpart. Indeed, the shape of a power spectrum of an *unbounded* time series explicitly depends on the statistics of the members and the length of the corresponding time series. The fundamental importance of the existence of time-series invariants will be best revealed in the next Chapter where we prove the additive separability of the power spectrum of a BIS to a specific discrete band (which represents the corresponding homeostatic pattern) and a noise band of universal shape. The value of this decomposition is that it renders constant accuracy of both occurrence and recording of a given homeostatic pattern. The major move ahead, developed at the end of Chapter 2, is that on the grounds of this separability we are able to work out early warning signs for any change in the homeostasis.

However, some very important topics still remain assumptions. First in this line comes the matter about the uniform contribution of all time scales. So far it has been taken for granted assuming that U-turns happen without involving any specific physical process and without destabilization of the system but without proving it. In Chapter 5 we will prove mathematically that the obvious relation between stretching and folding mechanism and the stability of the system prompts association of the issue about the scale-free execution of U-turns with

the scale-free properties of stretching and folding established in the present Chapter. Further in the line stands the assumption about the irregularity of the fluctuations succession, i.e. we post the question what drives the fluctuations to appear irregularly. We will come to these questions later in Chapter 3 and Chapter 5 of the book.

REFERENCES

Feller, W. (1970). *An introduction to probability theory and its applications*. New York, NY: John Wiley & Sons. doi:10.1063/1.3062516

Koleva, M. K. (2005). *Fluctuations and long-term stability: From coherence to chaos*. Retrieved from http://arxiv.org/abs/physics/0512078

Koleva, M. K., & Covachev, V. (2001). Common and different features between the behavior of the chaotic dynamical systems and the 1/f^alpha-type noise. *Fluctuation and Noise Letters, 1*(2), R131-R149. arXiv:cond-mat/0309418

APPENDIX

$$J(a,b) = \int\limits_{a}^{b} x^{\pm\nu(x)} dx. \tag{1}$$

$$J(a,b) = \frac{b^{1\pm\nu(b)}}{1\pm\nu(b)} - \frac{a^{1\pm\nu(a)}}{1\pm\nu(a)} + \lim_{\varepsilon\to 0} \sum_{k=1}^{[(b-a)/\varepsilon]} R_k, \tag{2}$$

$$R_k = (a+k\varepsilon)^{\pm\nu(a+(k-1)\varepsilon)+1} - (a+k\varepsilon)^{\pm\nu(a+k\varepsilon)+1} \approx (a+k\varepsilon)^{\pm\nu(a+k\varepsilon)+1}\left((a+k\varepsilon)^{-\delta_k\varepsilon} - 1\right),$$

$$\delta_k = \nu'(a+k\varepsilon).$$

Since it is supposed that $\nu(x)$ has finite everywhere derivatives, δ_k is finite for every k. In turn, this provides that $\left((a+k\varepsilon)^{-\delta_k\varepsilon} - 1\right) \to 0$ at $\varepsilon \to 0$. So each term $R_k = 0$ at $\varepsilon = 0$. Thus:

$$J(a,b) = \frac{b^{1\pm\nu(b)}}{1\pm\nu(b)} - \frac{a^{1\pm\nu(a)}}{1\pm\nu(a)}. \tag{3}$$

Let us now consider the case when $\nu(x)$ is a continuous function with finite everywhere derivatives such that it crosses 1 at the point $x = c$ and $c \in [a,b]$. Then:

$$J(a,b) = \lim_{\varepsilon\to 0}\left[\int\limits_{a}^{c-\varepsilon} x^{\pm\nu(x)} dx + \int\limits_{c+\varepsilon}^{b} x^{\pm\nu(x)} dx\right] = \frac{b^{1\pm\nu(b)}}{1\pm\nu(b)} - \frac{a^{1\pm\nu(a)}}{1\pm\nu(a)} + \lim_{\varepsilon\to 0}\left((c-\varepsilon)^{\pm\nu(c-\varepsilon)+1} - (c+\varepsilon)^{\pm\nu(c+\varepsilon)+1}\right). \tag{4}$$

Taking into account that:

$$\left| \nu\left(c - \varepsilon\right) \right| = 1 - \nu'\left(c\right)\varepsilon$$
$$\left| \nu\left(c + \varepsilon\right) \right| = 1 + \nu'\left(c\right)\varepsilon,$$

$$\lim_{\varepsilon \to 0}\left(\left(c - \varepsilon\right)^{\pm\nu\left(c-\varepsilon\right)+1} - \left(c + \varepsilon\right)^{\pm\nu\left(c+\varepsilon\right)+1}\right) \approx$$
$$\lim_{\varepsilon \to 0}\left(c^{-\nu'\left(c\right)\varepsilon} - c^{\nu'\left(c\right)\varepsilon}\right) \to c^0 - c^0 = 0.$$

Therefore again:

$$J\left(a, b\right) = \int_{a}^{b} x^{\pm\nu\left(x\right)}dx = \frac{b^{1\pm\nu\left(b\right)}}{1 \pm \nu\left(b\right)} - \frac{a^{1\pm\nu\left(a\right)}}{1 \pm \nu\left(a\right)}. \tag{5}$$

However, the result seems controversial because the derivative of $J\left(a, b\right)$ deviates from the *l.h.s.* of (5). Next a proof that this argument is alias and (5) holds whenever $\nu\left(x\right)$ is a continuous function with finite everywhere derivatives is presented.

To simplify the further calculations let us present (5) in a slightly modified form, namely:

$$J\left(a, b\right) = \int_{a}^{b}\left(\pm\nu\left(x\right)\right) x^{\pm\nu\left(x\right)-1}dx = b^{\pm\nu\left(b\right)} - a^{\pm\nu\left(a\right)} \tag{6}$$

$$x^{\pm\nu\left(x\right)} = \exp\left(\pm\nu\left(x\right)\ln\left(x\right)\right) \tag{7}$$

$$\frac{dx^{\pm\nu\left(x\right)}}{dx} = \pm\nu\left(x\right) x^{\pm\nu\left(x\right)-1} + \nu'\left(x\right)\ln\left(x\right) x^{\pm\nu\left(x\right)} \tag{8}$$

Obviously (8) deviates from the *l.h.s.* of (6) by the factor $\nu'\left(x\right)\ln\left(x\right) x^{\pm\nu\left(x\right)}$.

Let us now apply the rigorous definition of a derivative. In our case it reads:

$$\frac{dx^{\pm\nu(x)}}{dx} = \lim_{\varepsilon\to0} \frac{(x+\varepsilon)^{\pm\nu(x+\varepsilon)} - x^{\pm\nu(x)}}{\varepsilon} = \lim_{\varepsilon\to0} \frac{\exp\left(\pm\nu(x+\varepsilon)\ln(x+\varepsilon)\right) - \exp\left(\pm\nu(x)\ln(x)\right)}{\varepsilon} \tag{9}$$

For each x fixed and non-zero one can define the small parameter $\frac{\varepsilon}{x}$. Hence, the Taylor expansion of $\nu(x+\varepsilon)\ln(x+\varepsilon)$ to the first order reads:

$$v(x+\varepsilon)\ln(x+\varepsilon) = \left(v(x) + \varepsilon\nu'(x)\right)\left[\ln\left(1+\frac{\varepsilon}{x}\right) + \ln(x)\right] =$$
$$\left(\nu(x) + \varepsilon\nu'(x)\right)\left[\frac{\varepsilon}{x} + \ln x\right] = \nu(x)\ln(x) + \nu(x)\frac{\varepsilon}{x} + \varepsilon\nu'(x)\ln(x) \tag{10}$$

Then:

$$\exp\left(\pm\nu(x+\varepsilon)\ln(x+\varepsilon)\right) = x^{\pm\nu(x)} \exp\left(\pm\nu(x)\frac{\varepsilon}{x}\right)\exp\left(\varepsilon\nu'(x)\ln(x)\right) \tag{11}$$

The expansion of $\exp\left(\pm\nu(x)\frac{\varepsilon}{x}\right)$ to the first order of $\frac{\varepsilon}{x}$ is always available since by the limit $\varepsilon\to0$ in (9) is taken at fixed value of x. It yields $\left(1\pm\nu(x)\frac{\varepsilon}{x}\right)$. However, the expansion of the term $\exp\left(\varepsilon\nu'(x)\ln(x)\right)$ into series is incorrect because of the divergence of $\ln(x)$ at small x. Furthermore, x can always be made arbitrarily small by appropriate choice of the unites. That is why the meeting of the scaling invariance requires taking the limit $\varepsilon\to0$ at fixed but still finite x. In result this makes the term $\exp\left(\varepsilon\nu'(x)\ln(x)\right)$ equal to 1 at each value of x. So, (11) becomes:

$$\exp\left(\pm\nu(x+\varepsilon)\ln(x+\varepsilon)\right) = x^{\pm\nu(x)} \exp\left(\pm\nu(x)\frac{\varepsilon}{x}\right) \tag{12}$$

Note that this operation provides scaling-invariant form of the small parameter $\dfrac{\varepsilon}{x}$ in (11). Note also that the value of $\dfrac{\varepsilon}{x}$ does not depend on the choice of units! So it turns out that the derivation preserves the scaling-invariance!

Thus, (8) yields:

$$\frac{dx^{\pm\nu(x)}}{dx} = \pm\nu\left(x\right)x^{\pm\nu(x)-1} \tag{13}$$

So, (13) matches our result (6).

Chapter 2
Time Series Invariants under Boundedness:
Information Symbols and Early Warning Signs

ABSTRACT

One of the basic notions of any type intelligence expressed in a semantic-like manner is the notion of a letter (character). In view of the concept of boundedness, a letter must be implemented as characteristic of a specific natural process. This view sets one of the fundamental demands to every letter to be its autonomity, i.e. to retain its specific characteristics on reoccurrence. Thus, the matter of implementing of a letter turns apparently related to the issue about the robustness of homeostasis to small environmental fluctuations. It is proven that the response of a stable complex system is additively decomposable to a specific steady part and a universal one so that the steady part reoccurs with the same accuracy in an ever-changing environment. This makes the association of the notion of homeostasis with the steady pattern appropriate. In result, the corresponding characteristic of the homeostasis appears as a suitable candidate for a "letter".

INTRODUCTION

The empirical analysis of complex systems behavior makes it clear that even though they are very different in nature and each and every of them respond in a highly specific manner to the tiniest variations of the environment, their behavior shares the common property of persistent coexistence of specific and universal properties. This persistency poses the fundamental question whether behind it there is a common constraint embedded in the self-organization of the complex

DOI: 10.4018/978-1-4666-2202-9.ch002

systems. What makes the task enormously difficult is that the ubiquity of that separation is shared by an enormous diversity of the environmental variations and the corresponding response. This makes a lot of scientists to preclude any general rule able to govern the observed ubiquitous coexistence between specific and universal properties. To remind, this behavior has been observed for systems as diverse as quasar pulsation, heartbeat in mammals, currency exchange rates, DNA sequences etc. Contrary to this wide-spread understanding we put forward the idea that, though there is no general rule to govern the complex systems behavior, there do exist a general *operational protocol* that governs their behavior. Our reason for adopting this point of view is the clear indication that only *stable* complex systems retain common behavior. In turn, the constraint of stability naturally draws the limitations of our approach: it is available for stable complex systems only.

Our reason for highlighting the stability of complex systems as a key property for explanation of their common behavior is grounded on the assumption that the response of a complex system is a collective phenomenon whose self-organization explicitly involves another persistent property of complex systems: that of their multi-level hierarchical semantic-like self-organization. As it will be demonstrated throughout the book, the strategic aim of that hierarchy is to tighten the overall response of the corresponding system. This is implemented by means of diversifying the environmental impact onto different hierarchical levels so that the impact of one level on its neighbors to be kept permanently bounded. This explains how and why the "environment" is kept bounded for such diversity of scales and systems, i.e. it ranges from atomic to cosmic scales, from quantum systems to quasars. In the next Chapter we will demonstrate how the boundedness, set on quantum level, restrains local fluctuations of e.g. concentrations on the next level of hierarchy, to be bounded. Later, in Chapter 7, we will demonstrate that the boundedness of the impact

of one hierarchical level onto another is a generic property of the multi-level hierarchical semantic-like self-organization.

The major goal of semantic-like hierarchy is sustaining boundedness of each and every hierarchical level. In turn, as it will be proven in the section *Power Spectrum as a Letter* under the mild constraint of boundedness alone, the steady pattern of the response of a stable complex system, presented through a BIS, maintains its characteristics with constant accuracy which is insensitive to the statistics of the corresponding environmental impact. Consequently, this result gives not only a credible general explanation for the persistent coexistence of specific and universal properties, but substantiates the use of steady patterns as "letters", i.e. as building blocks of semantics. Later, in Chapter 4 and Chapter 7, we will demonstrate that the executed semantics is organized in a non-extensive way similar to that of a human language. A great value of this result is that it opens the door for an explicit formulation of "early warning signs" for development of any changes in the steady pattern far before any of them reaches full size. Moreover, the early warning signs explicitly discriminate between events with hazardous effects on a given system from "non-hazardous" ones.

It is worth noting that the considered in the present book hierarchical organization is different from the so far known types of hierarchy in the following sense: the introduced by us type of hierarchy is aimed towards tightening robustness of homeostasis and thus its semantic-like properties appear as product of this aim. To be specific, let us consider for example the major difference between our (semantic-like) hierarchy and the pyramidal one. The pyramidal type hierarchy aims towards appropriate detailed classification of properties and functions. As a result the pyramidal type hierarchy turns very vulnerable to small "errors" in any particular classification. On the contrary, the semantic-type hierarchy renders each and every inter- and intra-level feedbacks bounded. In turn, as we will demonstrate further, the latter provides robustness of semantic-like response to local failures. Still, the most exclusive property of semantic-like hierarchy is substantiated by the highly non-trivial interplay between an inter-level feedback and the self-organization of the level which it acts onto – the hallmark of that interplay is the appearance of a non-recursive extra component in the corresponding power spectrum. The consequences are far-going: (i) it opens the door to a unique opportunity for providing non-recursive computing, i.e. reaching algorithmically irreducible to one another symbols by means of finite number of steps implemented by corresponding physical processes; (ii) it renders the law of Nature unique and irreducible to recursive computing alone; (iii) it opens the door to a generic recipe how to "built-up" a semantic-like hierarchy of "recursive" units. To be specific, in the next Chapter we will provide the "recipe" how to produce a bounded inter-level feedback on quantum level, i.e. out of atoms and molecules whose interaction is well defined by recursive means. Later, in Chapter 10 we will discuss how networks of computers, each of which is a Turing machine, would look like as if their organization is semantic-like. The task of the present Chapter is to demonstrate how a non-recursive component appears in a power spectrum and its role in epistemology; we will also discuss the advantages of a non-recursive computing.

BACKGROUND

The general frame for implementing the idea of a self-regulated response is our concept of boundedness which asserts that any stable organization of matter, energy and information is created and maintained by involving and exchanging only bounded amount of energy and matter. Another item of our concept is the release of the environmental variations and the corresponding response from any global pre-determination other than the

mild assumption of boundedness; in other words, we set the response to be defined by local rules only. However, it seems that now we jump form one fundamental difficulty to another, namely: the release from global pre-determination makes us positive to assume that it puts a ban over any predictability of the future behavior. Indeed, this would be the case if we leave the impact arbitrary; then, it is to be expected that any hazardous for the system event would happen in a catastrophic-like way, i.e. when certain external parameters are fine-tuned to certain variables characterizing the system. So, the lack of any preliminary "signs" for a possible "disaster" regards the matter of predicting the future behavior crucially important. However, our hope is that the concept of boundedness fundamentally changes this paradigm since the boundedness of rates and of amplitudes introduces non-physical correlations among distant responses. In result the notification of early warning signs is to be expected. Which are these signs and how certain is their "reading" will be revealed in the section *Time Series Invariants and Early Warning.*

The non-triviality of the matter about early warning is best revealed by the following puzzle: on the one hand, the idea of predictability through early warning seems inherently compromised by the established in the previous Chapter result that one of the time-series invariants of a bounded time series (BIS), K-entropy, sets universal constraint over the predictability of both the nearest and any other future behavior. On the other hand, the universality of the K-entropy could be viewed as confirmation of the locality of the response in the sense that the local response is specified by only current impact and current state of the system, i.e. it supports our assumption about lack of any pre-determination of the corresponding environmental impact. The key property of K-entropy which will help revealing the puzzle is that, for bounded sequences, K-entropy is always finite on the contrary to its infinity as it is for random but *unbounded* sequences.

Another obstacle to the idea of early "warning" criterion comes from our next time-series invariant: the topological equivalence between the parameterization of the phase space through angle and through the duration of the corresponding excursion. One of the consequences of this equivalence is that the non-physical correlations are robust to the coarse-graining. Then, is it ever possible to make a distinction between them and any sign of early "warning"?! To remind, that the coarse-graining is mathematical operation that acts non-linearly and non-homogeneously over the members of a bounded irregular sequence (BIS). The meaning behind the coarse-graining is that it accounts both for the locality of the response and for the imperfection of its recording. The lack of any global pre-determination of the response renders any long-run adjustment of the recording equipment impossible. Thus, the catastrophe of our theory seems complete: our ambition to renounce any global pre-determination seems to bring us to a deadlock – it is impossible not only to predict nearest future but it turns impossible even to make a record!!! However, our story has a happy ending which is provided by the use of the last of out time-series invariants, the power spectrum and its highly non-trivial compatibility with the other time series invariants. You can find out how all this happens in the present Chapter.

At first glance it seems that we step aside from the main idea of the book which is how the boundedness gives rise to self-organized semantics. Next we will demonstrate that the issues about the stability and predictability are an indispensable part of the systematic theory about how the boundedness brings about self-organized semantics.

We have already pointed out that both intelligent and non-intelligent systems belong to the same class of complex systems. Thus, it is natural to suggest that the intelligence is executed by natural processes. Further, in this line of reasoning, the suggestion that the multi-level hierarchical-like organization plays a crucial role for a non-

extensive organization of the intelligence appears. Since we already mention that the inter-level feedbacks act in such a way so that to tighten the response and thus to increase the stability of a system, it is natural to suggest that the multi-level hierarchical-like organization and its properties are the implement for exerting intelligence. Then, it is to be expected that the properties of the self-organization and its functioning are matched by the properties, organization and execution of the corresponding intelligence which we believe is organized and executed in a semantic-like manner.

One of the basic notions of any type intelligence expressed in a semantic-like manner is the notion of a letter (character). In view of the above setting, a letter must be implemented as characteristic of a specific natural process. This view sets one of the fundamental demands to every letter to be its autonomy, i.e. it must retain its specific characteristics on reoccurrence. Thus the matter of implementing of a letter turns out to be related to the issue about the robustness of homeostasis to small environmental fluctuations. Thus, we come back to the issue of separation of the properties to a universal and a specific part and to the view that the specific part is a characteristic of the homeostasis. Now we make our assumption more precise by suggesting further that the response of a stable complex system is a cooperative phenomenon where the intra-level response is organized as a pattern whose local stability is sustained by inter-level feedbacks. In turn, this suggestion opens the door for association of the intra-level pattern with the homeostasis while the corresponding inter-level feedback substantiates the universal part. The success of the assumption will be the proof that the characteristics of the homeostasis are robust to small environmental variations and remain the same under reoccurrence. In result, the corresponding characteristic of the homeostasis will stand up as an appropriate candidate for a "letter".

Even though the association of a letter with certain characteristics of homeostasis has its ample understanding, it still remains unclear what type of intelligence is to be expected. The problem is that on the one hand, we are not used to associate the semantic-type intelligence with specific physical processes. On the other hand, the fine regulation of the multi-level hierarchy of complex systems suggests that corresponding "semantics" has its specific "price" in energy and matter. Thus, it appears that thinking is a "hard work" and the intelligence turns as a measure of this "hard work". But, how to compare one "hard work" with another so that to distinguish one semantic meaning from another.

Further, is it possible to compute that "hard work"? This is a question of fundamental importance since it opens the door to the discussion whether both the intelligence and the Universe are Turing machines and thus are they computable in the traditional sense? This discussion will proceed throughout the entire book since our major goal is to demonstrate that the self-organized semantics executed by intelligent complex systems is irreducible to a Turing machine. The contribution of the present chapter will be to demonstrate that the association of a "letter" with certain properties of the power spectrum, one of the time-series invariants of a BIS, fundamentally expands its power because now the "letter" appears not only as an information unit but as a semantic unit as well. It should be stressed on the new role of that information unit. Unlike the traditional approach where the computing is executed in an intensive way by linear processes only and the information units are indifferent to the "distance" between them, now the information symbol has a two-fold meaning whose major property is that accounts not only for the current state but comprises the distance to the admissible ones as well. Its two-fold presentation implies that: (i) on the one hand it appears as the measure for energy/matter involved for creating the information state; (ii) on the other hand, viewed as an ordered set of members, it stands for the details of the structure of the corresponding homeostatic pattern. What constitutes the

essential difference between these counterparts is that, while the matter/energy involved and necessary to reach another information unit is always finite, the "distance" in the information is always non-recursive since the characteristic of each information unit comprises specific irrational number(s). Thus, the algorithmic irreducibility of the information "distance" and energy/matter involved for its covering put the grounds for non-extensive computing and for constituting semantics which is not reducible to formal logic.

Outlining, the major distinction between the traditional algorithmic approach and ours is that, unlike the former where the information is the measure how far a given string of symbols deviates from a random sequence, in the present approach the information is associated with specific characteristics of the multi-level hierarchical-like organization and is expressed in a non-recursive manner.

POWER SPECTRUM AS A "LETTER"

A key point in the development of a systematic theory of self-organized semantics is establishing the appropriate characterization of the corresponding alphabet, grammar and semantic rules. An obvious property of the alphabet is that the recognition of a given "letter" in different words implies that it retains the same characteristics on reoccurrence and that they are unique in the sense that these characteristics must be algorithmically non-reachable from any other letter. It is important to stress that the algorithmic non-reachability of any two letters is not reducible to the physical distance between them measured by the energy/matter involved in the transition. The ban over any algorithmic reachability of one letter from another implies that the information content of each of them could not be reduced to the information content of any other by means of formal logic operations.

To remind that the formal logic "transforms" a set of natural numbers (input) to another set of natural numbers (output) according to a specific set of rules (algorithm, program). The physical implement of exerting that transformation is indifferent to the algorithm if only it is set to execute linear processes only. Therefore, it is obvious that when the hardware is "linear", the information appears as a quantity which measures the "switches" from one information symbol to another; put it in its most popular form – it measures the quantity of bits. However, the crucial drawback of that setting is that the quantity of bits tells noting about the "quality" of information, i.e. it tells noting about its meaning and it is even silent to whether it comes from a random sequence. Thus, the "linear", i.e. algorithmic information has not autonomous comprehending; it could be extracted by an 'external mind" only.

The major step on our road to overcome the problem with the autonomous comprehending is the assumption about separation of algorithmic and physical characterization of the letters. Thus we assert that: (i) the information symbol (letter) must be algorithmically non-reachable from any other; (ii) at the same time each letter must be physically reachable from the other letters. This is the first step on the road to self-organized semantics whose major property is expected to be its autonomous comprehending. The goal of the present Chapter is to delineate the implements of a non-recursive characterization of an information symbol. We start with its most immediate properties: its robustness and its non-recursive characterization.

In the setting where the semantics is supposed to be executed by natural processes it is obvious that we must start with the most general frame within which we will consider its properties. The first property stems from the observation that the natural intelligence is executed in an ever-changing environment so that its major characteristics remain the same. Thus, both in a quite weather and in a storm our alphabet is the same. This observation suggests that the letters are defined in an interval of external parameters within which they stay intact. On the other hand, the question how to work it together relate it with the observation that the response is separable to a specific and universal part still stands. We already have suggested that the universal part is brought about by the inter-level feedbacks whose purpose is to strengthen the response and thus to make the system more stable while the specific one comes out from the specific intra-level self-organized pattern that represents homeostasis.

The task of this section is to prove that a non-recursive characterization of an information symbol (letter) is available by means of one of the time-series invariants of a BIS considered in the previous Chapter. For the successful definition of this type of characterization of an information symbol we have to prove that: (i) there exists certain characteristics that reoccur with the same accuracy even though the environment does *not* re-occur exactly the same; (ii) these characteristics are covariant, i.e. they neither depend nor introduce any special reference point in the time; (ii) these characteristics comprise a specific non-recursive component.

The first task is grounded on considering the first of the time series invariants: the power spectrum. In the previous Chapter we have established that the power spectrum of a zero-mean bounded irregular sequence (BIS) fits the shape $1/f^{\alpha(f)}$ where $\alpha(f)$ is linear function. The proof of that shape is grounded on two mild constraints only: the amplitude of the members is bounded and no physical processes are initiated by executing "U-turns" on reaching the thresholds of stability. Note, that the proof does not involve any specification of the distribution or any other particularities of the members. Further, the shape is proven invariant under coarse-graining. To remind, the operation of coarse-graining is a mathematical operation which acts on the members of a BIS in non-linear and non-homogeneous way under the single condition of preserving the boundedness of each member. The invariance of the shape to the coarse-graining implies that the "recipients" of that signal gain the same knowledge irrespectively to the current status of their senses and adjustment abilities of the latter.

Outlining, the insensitivity of the shape of the power spectrum of a zero-mean BIS to the statistics of the environment and the properties of the recipient implies that it plays the same robust role in the power spectrum. As it will be proven next, this property provides constant accuracy for appearance of the other component of a power spectrum which is the corresponding specific discrete band. Put in other words, we will prove that the boundedness alone is enough for providing both constant certainty of our knowledge and robust existence in an ever-changing environment of the discrete band in each and every BIS. This makes a discrete band "identifiable" and thus justifies suggestion that a "letter" is to be associated with the discrete band of the power spectrum Next we have to prove that the mild constraint of boundedness alone is enough for providing additive decomposition of the power spectrum of BIS to a specific discrete band and universal noise band. Moreover, the interplay between the obtained additivity and the exclusive properties of the universal band of a BIS renders constant accuracy for reoccurrence of the discrete band. We consider the notion of information as possibility for obtaining certain knowledge about identification of an object from any other in an ever-changing environment. That is why we put the proof in the general frame of the discussion about separation of genuine information from accompanied noise.

A key problem of processing information is outlining the general conditions that provide the separation of the genuine information from the accompanied noise. Here this problem arises from our assumption that a steady state in an ever-changing environment is represented by bounded irregular sequences (BIS) among which are sequences that comprise both discrete bands (information) and continuous bands (noise). It is obvious that the reproducibility of the information is met only if the accuracy of extracting information is independent from the statistics of the noise. The ontological aspect of the problem is that the target independence is an indispensable part of preventing the interference between the causal relations that bring about information and the noise. Indeed, note that while the information remains the same on repeating the same external stimuli, the noise realization is ever different. The task of this section is to demonstrate that the problem is successfully resolved in the frame of the proposed by us concept of boundedness.

Let us consider a BIS which comprises both a discrete and a noise band in its power spectrum. We will prove that the discrete and the noise band of every BIS are additively superimposed; moreover the shape of the noise band is insensitive to the statistics of the "noise" (environmental variations) and to the status of the recipient. More precisely, it could be only the already established $1/f^{\alpha(f)}$ where $\alpha(f)$ is linear function.

Another important aspect, related to the idea of information viewed as a measure of causality, is that the additivity of the noise and the discrete band along with the insensitivity of the shape of the noise band to the statistics of the noise are the necessary and sufficient conditions for preventing the interference between causality and noise. So, indeed the boundedness is the target general condition that provides reproducible separation of the genuine information from the noise. But now we shall prove more: we shall prove that the only decomposition of the power spectrum compatible with the boundedness is the additivity of the discrete and continuous band. To prove that assertion let us suppose the opposite, namely: suppose that the discrete and the continuous band are multiplicatively superimposed:

$$A(t) \propto \int_{f_0}^{\infty} g(f) \exp(ift) df \sum_{l=1}^{\infty} c_l \exp(ilt)$$

$$(2.1)$$

where $A(t)$ is the amplitude of the variations in the time series; $g(f)$ are the Fourier components of the continuous band and c_l are the Fourier components of the discrete band. The purpose for representing the multiplicative superposition of the discrete and continuous band through their Fourier transforms is that it makes apparent the permanent violation of the boundedness of $A(t)$ even when both the noise and the information are bounded. Indeed, it is obvious that there is at least one resonance in every moment regardless to the particularities of the causal relations between c_l and the noise $g(f)$. In turn, the resonances break the boundedness of $A(t)$ because they make the local amplitude to tend to infinity. This, however, violates energy conservation law because it implies that finite amount of energy/substance (bounded information and bounded noise) generates concentration of infinite amount of energy/substance (infinitely large amplitude of fluctuations at resonance).

However, it is worth noting that alone the additive superposition of a discrete and a continuous band is not enough to provide the reproducibility of the information. Indeed, the additive superposition reads:

$$A(t) \propto \int_{f_0}^{\infty} g(f) \exp(ift) df + \sum_{l=1}^{\infty} c_l \exp(ilt)$$

$$(2.2)$$

It is obvious that the reproducibility of the genuine information encapsulated in c_l requires insensitivity of $g(f)$ to any noise realization, i.e. to the noise statistics. As we have already proved

in the previous Chapter, this is possible only for bounded noise; then the continuous band in its power spectrum fits the same shape regardless to the particularities of the noise statistics; and as we have already proven this shape is $1/f^{\alpha(f)}$.

Summarizing, we can conclude that the boundedness is that single general constraint which provides preventing interference between causality and noise in a self-consistent way. The key importance of that result is its immense ubiquity. Indeed, since the boundedness as the constraint that prevents interference between causality and noise is independent from any particularities of the processes that generate time series, it is equally available for systems of different nature ranging from cosmic rays to DNA sequences and financial time series to mention a few.

The permanent maintenance of constant accuracy of the discrete band reoccurrence along with the lack of any other special time scales (frequencies) signaled out by the noise allows to conclude that: the system stays stable in a co-variant way, i.e. its behavior is neither associated with any reference point in the time nor it selects any such. This conclusion gains additional support by the proof made in previous Chapter that the variance of a zero-mean BIS is constant and depends neither on the length of the time series nor on its position on the time scale. Keeping in mind that the variance of the power spectrum is measure for the energy/matter exchanged with the environment, its constancy decisively suggests that the system is in a stable steady state which is in balance with the environment.

In turn, the covariance of the discrete part of the power spectrum allows us to consider the causality of the homeostasis as specific law that is grounded on the ordered relations among the frequency members in the discrete part. Yet, the distinctive property of that law is that unlike the traditional laws it must be expressed in a non-recursive way because of the presence of another, non-recursive frequency in the power spectrum.

We have already mentioned that the presence of a non-recursive frequency is crucial for making the corresponding "letters" algorithmically unreachable one from another.

WHY DO WE NEED NON-RECURSIVE FREQUENCIES AND WHERE THEY COME FROM?

Why Do We Need Non-Recursive Information?

One of the most fundamental problems in the information theory is to what extend the information and its processing is related to the organization and functioning of the natural processes. The enormous development of the information theory culminates in the recently widely discussed matter whether the laws of the Universe are computable. The extreme importance of that matter is provoked by the universality of Turing machines which comes to say that every law, viewed as a quantified relation between certain variables, expressed in a closed form, is computable. Then, an immediate outcome of the traditional algorithmic theory is that the variety of laws turns compressible (reducible) to its shortest form so that there is no shorter algorithm executable on any Turing machine. In turn, the existence of universal algorithm (shortest program) whose major property is its insensitivity to the hardware (Turing machine) implies that our Universe is subject to universal law so that each of it's constituents obeys it irrespectively to its specific properties. Thus the algorithmic compressibility smears out any difference that sets apart one specific property from another.

Although the idea of a universal law has been appealing during the times, it still remains contra intuitive because we do "feel" the difference between specificity of different events and phenomena as well as their uniqueness and their irreducibility to one another. Then a closer cold-minded look on the idea of a universal law displays its inherent controversy. The first one in the line looks like a technical one: how to compress a stochastic sequence where the response is released from any pre-determination other than boundedness? As we have already demonstrated in the previous Chapter there exists a universal ban over any predictability of the nearest and any distant behavior of every complex system. Consequently, the ban over predictability puts a ban on any compressibility in its traditional understanding.

The second controversy commences from the existence of emergent phenomena. Although we are used to associate the emergent phenomena with highly complex systems, actually emergent properties and phenomena occur at every level of matter/energy organization. Thus, an example for emergent property is the glitter of some metals: not a hint of a glitter is embedded in the Hamiltonian of any formal description of a metal. Another example for emergent phenomenon is the emergence of wave-like behavior in Belousov-Zhabotinsky reaction. The self-organized pattern emerges spontaneously from a homogeneous solution. What is important with it is that the spontaneous pattern formation implies anti-thermodynamic behavior, i.e. the emergence goes from a state of maximum entropy (homogeneous solution) to a state of lower entropy (self-organized pattern). Therefore, the spontaneous pattern formation violates the Second Law. Note, however that this violation happens only when external parameters are fine-tuned to certain specific for the solution variables. Thus, once the Second Law holds, the next time it turns violated. Then, how the violation of one and the same law along with its validity would be incorporated in a single Universal law ?!

In addition, what is the most counter-intuitive is that an universal law should comprise not only both the violation and the validity of the Second Law, but the glitter of the gold along with the structure and functioning of our mind because it is made of the same atoms and molecules as other objects in the Nature and thus it must be subject to the same law. The latter brings about a vicious circle: it is well known that the "decoding" of any information in the traditional information theory is implemented by an "external mind" which is practically our human mind. Thus, putting it on the same ground as the subjects that it is supposed to decode yields impasse – who then would decode the information independently from the subjects of decoding. Thus, even though we supposed a universal law to exist, we are not able to recognize it.

Next in the line of controversies comes the following: even if we claim the universal law unrecognizable and reduce the task of science as finding out "partial laws", how could we recognize improvable statements and how to distinguish a true law. This matter is of immense methodological importance in the study of new phenomena. What experienced scientists know is that we judge any new observation in the context of our current knowledge on the matter. Then, depending on how well the observation fits the entire picture, we classified it as important or not. Yet, sometimes tiny deviations could overturn the things upside down. We have already encountered such critical reading of the one of the most ubiquitous and still mysterious phenomena – the famous $1/f$ noise.

The great mystery of the phenomenon $1/f$ noise consists of the following observations:

1. It is one of the most ubiquitous phenomena ever observed and it is an indispensable property of all complex systems;
2. It puzzles with the antagonism between the apparent stability of all systems where it has been observed and the divergence of their $1/f$ power spectrum.

To make this clearer, let us remind that the major characteristic of the $1/f$ noise is the presence of a continuous band in the power spectrum so that its infrared end fits the shape $1/f$ where f is the frequency. An exclusive property of that

shape is that the fit does not depend on the length of the time series. In turn, this implies that energy/matter exchanged with the environment increases with the length of the time series. Indeed, the exchanged energy is measured by:

$$\int_{1/T}^{A} \frac{1}{f} df \propto \ln A - \ln\left(\frac{1}{T}\right) = \ln A + \ln T \quad (2.3)$$

where T is the length of the time series and A is a characteristic point of the power spectrum where it is believed that a crossover to another fit occurs.

The conundrum arises from the reading of the power spectrum shape: on the one hand, the persistency of the shape and its ubiquity apparently suggests certain universality; on the other hand, the reading that the universality is to be associated with the infrared end only brings about its rejection: indeed, as one can see from Equation (2.3) every system should select a specific point in the time scale, which is counterpart to the frequency A in (2.3) at which the crossover to other slope of the power spectrum occurs. However, if such point would exist, it should be associated with some specific for the system process whose presence in the power spectrum would be denoted by a specific line with firm position in it. Not even a hint of such behavior is observed! Then, the avoidance of this problem seems to be accomplished by means of suggesting a single fit for the entire continuous band.

However, it seems that even the suggestion about a single fit for the continuous band is of no help because it does not brings about elimination of the divergence of the integral over the power spectrum. The importance of this integral is that it measures the exchanged with the environment energy/matter in any given time interval. So, its divergence implies that the amount of energy/matter exchanged with the environment gradually increases with the time interval. In turn, this divergence is completely incomprehensible in view of the observed stability and inalterability of the complex systems. Indeed, the assumption that the shape is not exactly $1/f$ but $1/f^{\alpha}$ where α varies from system to system but stays confined around unity, $\alpha \in (0,2)$, leaves the integral over the power spectrum divergent: it suffers either from an infrared catastrophe, or of an ultraviolet one: now the integral over the power spectrum reads:

$$\int_{1/T}^{T} \frac{1}{f^{\alpha}} df = T^{-\alpha+1} - \left(\frac{1}{T}\right)^{-\alpha+1} \tag{2.4}$$

However, the setting where we allow the shape of the continuous band in the power spectrum to vary, though in a very restricted way, from one system to another gives rise to the question about additional classification of systems according to whether they exhibit infrared or ultraviolet divergence. The positive answer to this question implies a departure for the concept of universality. Yet, any retreat from the concept of universality encounters its apparent empirical persistence thus posing the following question: why any such system stays stable regardless to whether it belongs to the "infrared" class or to the "ultraviolet" one.

Thus, we come to the conclusion that indeed, a tiny "shift" in the reading of an experiment could have crucial impact on the understanding of that "shift" and its place in the context of this or that general setting.

We hope that this brief excurse is a thoughtful example about the necessity of being cautious when judging whether a given observation fits a given paradigm or it appears a sign of a new one. In the next sub-section we will consider how another look on the $1/f$ noise phenomenon brings about not only convergence of the power spectrum but opens the door to non-recursive frequencies in the discrete part.

Metrics and $1/f^{\alpha(f)}$ Noise

Our contribution to the problem is to consider the phenomenon $1/f$ noise in the larger context of complex systems, namely as one of their common properties so that its explanation to be consistent with the explanation of the other properties. What we put foremost is the stability and the covariance of those properties, i.e. all common properties of the complex systems are exerted so that the corresponding system stays long-term stable and this stability neither involves nor selects any special time point. Thus, this observation prompts the suggestion that no special physical process is initiated by the $1/f$ noise. Further, we put forward the general assumption about the boundedness of the response and the release of its pre-determination. Now, we can specify our idea about ban over initiation of any physical process by the $1/f^{\alpha}$ noise by asserting that this ban implies that no "U-turn" is driven by a specific physical process. Consequently, the permanency of this ban suggests that the continuous band in the power spectrum fits a single universal type of shape which must such as to provide the convergence of the integral over it.

Now we face the major question: is there any fit that can eliminate the divergence of the integral over the power spectrum? We have proposed the fit $1/f^{\alpha(f)}$ where $\alpha(f)$ is a continuous monotonic function with two properties: it always approaches unity at its infrared end; it provides that the power spectrum does not signal out any special frequency. At first sight, it seems that this assumption is meaningless because it appears that it not only does not contribute anything to resolving the problem with the divergence but is a complete non-sense since it assigns different dimensionality to the successive frequencies in the corresponding band. The latter argument is easy to discard by pointing out that the shape $1/f^{\alpha(f)}$

is applied to the dimensionless form of the continuous band. This can be worked out non-ambiguously since we supposed that the $1/f^{\alpha(f)}$ noise is not to be associated with any special physical process.

Yet, the elimination of the power spectrum divergence is a highly non-trivial task. We already considered it in the Appendix to the Chapter 1 where the rigorous proof is presented in details. We wish to highlight two major points in that poof because of their fundamental connection with our present and further considerations. The first item considers the role of the continuity of the function $\alpha(f)$. To remind, we suggest that the $1/f^{\alpha(f)}$ noise band does not signals out any special frequency. Then, it is to be expected that the continuity of the function $\alpha(f)$ provides uniform contribution of each member in the noise band. Thus the frequency $f = \dfrac{1}{T}$ must contribute the same way as that of $f = \dfrac{1}{T+\Delta}$ where T is the length of the time series and Δ is small number. But, the uniform contribution to the integral over the noise band seems impossible since:

$$\int \frac{1}{f} df = \ln f \tag{2.5}$$

while

$$\int \frac{1}{f^{1\pm\Delta}} df = f^{\mp\Delta} \tag{2.6}$$

where Δ is arbitrary small number. Thus, it appears that the uniform contribution of all members of the noise band to the integral is fundamentally banned by the rules of the ordinary calculus. However, a closer look on the proof about elimi-

nation of that problem, presented in the Appendix to Chapter 1, show that this ban does not hold for the spaces with uniform metrics. Indeed, lets us present again the second part of the proof where the role of metrics is apparently utilized. In order to make this very important difference clear in full details next we reproduce a part of the poof given in the Appendix to Chapter 1:

To simplify the further calculations let us present (5) in a slightly modified form, namely:

$$J(a,b) = \int_a^b \left(\pm\nu(x)\right) x^{\pm\nu(x)-1} dx = b^{\pm\nu(b)} - a^{\pm\nu(a)} \tag{6}$$

Indeed, the application of the formal differentiating rules to the function $x^{\pm\nu(x)}$ in its exponential presentation:

$$x^{\pm\nu(x)} = \exp\left(\pm\nu(x)\ln(x)\right) \tag{7}$$

yields:

$$\frac{dx^{\pm\nu(x)}}{dx} = \pm\nu(x) x^{\pm\nu(x)-1} + \nu'(x)\ln(x) x^{\pm\nu(x)} \tag{8}$$

Obviously (8) deviates from the l.h.s. of (6) by the factor $\nu'(x)\ln(x) x^{\pm\nu(x)}$.

Let us now apply the rigorous definition of a derivative. In our case it reads:

$$\frac{dx^{\pm\nu(x)}}{dx} =$$
$$\lim_{\varepsilon\to 0} \frac{(x+\varepsilon)^{\pm\nu(x+\varepsilon)} - x^{\pm\nu(x)}}{\varepsilon} =$$
$$\lim_{\varepsilon\to 0} \frac{\exp\left(\pm\nu(x+\varepsilon)\ln(x+\varepsilon)\right) - \exp\left(\pm\nu(x)\ln(x)\right)}{\varepsilon} \tag{9}$$

For each x fixed and non-zero one can define the small parameter $\dfrac{\varepsilon}{x}$. Hence, the Taylor expansion of $\nu\left(x+\varepsilon\right)\ln\left(x+\varepsilon\right)$ to the first order reads:

$$
\begin{aligned}
v\left(x+\varepsilon\right)\ln\left(x+\varepsilon\right) &= \\
\left(v\left(x\right)+\varepsilon\nu'\left(x\right)\right)&\left[\ln\left(1+\frac{\varepsilon}{x}\right)+\ln\left(x\right)\right] = \\
= \left(\nu\left(x\right)+\varepsilon\nu'\left(x\right)\right)&\left(\frac{\varepsilon}{x}+\ln x\right) = \\
\nu\left(x\right)\ln\left(x\right)+\nu\left(x\right)&\frac{\varepsilon}{x}+\varepsilon\nu'\left(x\right)\ln\left(x\right)
\end{aligned}
\tag{10}
$$

Then:

$$
\begin{aligned}
\exp\left(\pm\nu\left(x+\varepsilon\right)\ln\left(x+\varepsilon\right)\right) = \\
x^{\pm\nu\left(x\right)}\exp\left(\pm\nu\left(x\right)\frac{\varepsilon}{x}\right)\exp\left(\varepsilon\nu'\left(x\right)\ln\left(x\right)\right)
\end{aligned}
\tag{11}
$$

The expansion of $\exp\left(\pm\nu\left(x\right)\dfrac{\varepsilon}{x}\right)$ *to the first order of* $\dfrac{\varepsilon}{x}$ *is always available since by the limit* $\varepsilon \to 0$ *in (9) is taken at fixed value of* x. *It yields* $\left(1\pm\nu\left(x\right)\dfrac{\varepsilon}{x}\right)$. *However, the expansion of the term* $\exp\left(\varepsilon\nu'\left(x\right)\ln\left(x\right)\right)$ *into series is incorrect because of the divergence of* $\ln\left(x\right)$ *at small* x. *Furthermore,* x *can always be made arbitrarily small by appropriate choice of the unites. That is why the meeting of the scaling invariance requires taking the limit* $\varepsilon \to 0$ *at fixed but still finite* x. *In result this makes the term* $\exp\left(\varepsilon\nu'\left(x\right)\ln\left(x\right)\right)$ *equal to* 1 *at each value of* x. *So, (11) becomes:*

$$
\exp\left(\pm\nu\left(x+\varepsilon\right)\ln\left(x+\varepsilon\right)\right)=x^{\pm\nu\left(x\right)}\exp\left(\pm\nu\left(x\right)\frac{\varepsilon}{x}\right)
\tag{12}
$$

Note that this operation provides scaling-invariant form of the small parameter $\dfrac{\varepsilon}{x}$ *in (11).*

Note also that the value of $\dfrac{\varepsilon}{x}$ *does not depend on the choice of units! So it turns out that the derivation preserves the scaling-invariance!*

Thus, (8) yields:

$$
\frac{dx^{\pm\nu\left(x\right)}}{dx}=\pm\nu\left(x\right)x^{\pm\nu\left(x\right)-1}
\tag{13}
$$

So, (13) matches our result (6)."

The considerations that lead from ((11) to (12), apparently utilize the order in the execution of limits $\varepsilon \to 0$ and $x \to 0$ as a necessary condition for achieving uniform contribution of all members in an $1/f^{\alpha(f)}$ noise band, the most infrared included. And what sets this order is the assumption that the partitioning to sub-intervals does not change the metrics. The property of preserving the original metrics is viewed as the minimal condition for a sub-interval to preserve the same properties as the original interval. An immediate consequence of preserving the metrics, is that it bounds both ε and x to be expressed in the same units even when they participate in different mathematical operations such as taking the limits one after another; alongside, the lack of signaling out any special point sets the order of the limits: first, the limit $\varepsilon \to 0$ is executed and after that the limit $x \to 0$. In turn, the setting of the limits order substantiates the idea of continuity which results in dreamed convergence of the power spectrum as it has been proven in Chapter 1.

Thus, the most puzzling properties of $1/f$ noise phenomenon are finally revealed. Yet, the above considerations demonstrate that this happens only for systems where the response acquires certain metrics.

Metrics, $1/f^{\alpha(f)}$ Noise, and Non-Recursive Frequencies

Now we are ready to come back to the question about non-recursive characterization of the information symbols. Let us remind again that a non-recursive characterization serves as an implement for putting a ban over algorithmic reducibility of one information symbol to another. The key role of the ban is that the algorithmic irreducibility renders the process of compressing information not universal. In turn the latter guarantees diversity of laws of energy/matter organization in the Universe by making each of them unique. Yet, the essentially novel point is that the present setting considers the natural laws as steady relations which characterize self-organized response in a variable environment. Moreover, an essential ingredient of this setting is the assumption that the diversity is achieved through building of non-extensive multi-level hierarchical-like organization so that the inter-level feedbacks go both bottoms up and top down.

Then, it is to be expected that the novel setting would manifest the highly non-trivial role played by the multi-level hierarchical self-organization as an indispensable part of every specific law in a unique way. Thus our present task is to prove that the desired manifestation is substantiated through presence of a non-recursive frequency in every power spectrum which retains non-zero discrete part.

Initially, we put forward the idea that inter-level feedbacks serve as implements for tightening the local response and thus to strengthen the stability. Next in this line comes the proof that the accuracy of reproducing the discrete band in the power spectrum is sustained permanently constant even though the environment is not exactly reproduced. And since the discrete band in the power spectrum characterizes a specific steady pattern whose organization can be considered as a specific law, it seems this is enough. However,

here we encounter a very serious obstacle: it is well known that there is not such mathematical object the power spectrum of whose solution comprises any non-recursive frequencies.

If we proceed this consideration further, the bewildering becomes even more thought-provoking by taking into account the existence of metrics of the noise part of the response: on the one hand, it is to be expected that the metric characteristics of the noise part of the response such as the threshold of stability and the thresholds over the rates should participate in the power spectrum. Yet, their participation seems not to bring the desired result in view that the metric relations between the sizes of the corresponding thresholds and the characteristic sizes of the pattern are not bound to be non-recursive while our desire is their participation to be executed in a non-recursive manner.

Yet, a closer look shows that the above obstacles appear antagonistic only in the setting where the homeostatic and the noise part of the response are considered independent from one another. At first sight it seems correct to consider both parts separately since the one of them (the discrete) is highly specific while the other one (the noise) retains universality. In addition they are additively separated in the power spectrum as proven in the sub-section *Why Do We Need Non-Recursive Information?*

On the other hand, our assumption about setting of the properties of the response concerns only withdrawal of any global pre-determination. Thus, it is completely plausible to consider the local response as to be highly specific to the current impact and current state of the system. Thus, it is most likely to assume as its general property to be local inseparability to a specific and a universal part. Then, shall we consider that the entire response goes through self-organization of a specific pattern?! Yet, it seems that this way we again drive to nowhere because it is well known that the mathematical description of self-organized patterns is given by a system of autonomous ordinary or partial differential equations whose major property is that the power spectra of their solutions do not comprise either a noise band and/or irrational frequencies.

The rescue comes from our assumption that the universal part of the response comes from the inter-level feedbacks of the hierarchical organization. For being successful, this assumption must answer the following questions: how it is possible to comprise in a self-consistent way the high specificity of the local response and its separation to a universal and a specific part; the high non-triviality of the problem is set by the facts that on the one hand, this interplay should be non-linear in the solution while, at the same time, its power spectrum must be additively decomposable to a discrete and a continuous band. And the most important, how non-recursive characteristics are to appear?! Yet, the major question is about the mathematical implementation of our assumption. The suggestion about the specificity of the local response sets that the response should be defined mathematically by the same variables both for the specific and the universal part. The positive direction of this step is that in this way it provides the same metrics for the specific and the universal part of the response. Moreover, the common metrics is a necessary condition for any interplay between the discrete and the noise band. Thus, the common metrics opens the door to the desired non-recursive characterization which is expected to commence from such interplay.

The next step consists of defining new mathematical object, called by us bounded stochastic differential equations, in terms of which the evolution of a bounded response is considered. The study of these equations is postponed to Chapter 4, where it will be considered how the inter-level feedback yields their constitution and will be proven that their major property is the separation of the power spectra of their solutions to a discrete band and a continuous one of the shape $1/f^{\alpha(f)}$.

Then it will become apparent that interplay be-

tween the discrete and the noise band signals out any special frequency in non-recursive way.

To make clearer such an important topic we will consider in details how this happens. Let us take for granted that a given solution characterizes a self-organized pattern; thus the corresponding discrete part in its power spectrum comprises only rational members. Let us now consider the motion in the state space: in order to meet the boundedness, the solution "twists and turns" wildly around the "skeleton" "built" on the discrete part so that the deviations from the "skeleton" to stay within the corresponding thresholds; and since the deviations are irregular and thus the motion never closes, the motion on the "dressed" skeleton is equivalent to motion on a torus with irrational frequencies. Thus, though the thresholds of stability are free to be either rational or irrational, their match in the power spectrum is always through a non-recursive frequency.

Thus, we can conclude that indeed algorithmic irreducibility of one specific law to any other is feasible only for systems that are subject to multi-level hierarchical self-organization. This outlines a completely novel frame for comprehending the Nature organization and functioning. The traditional reductionist approach considers the environment specified and/or kept intact while the present one views the environment as ever-changing in a non-specified manner. Further, the traditional approach to the complexity reduces the Nature to separate levels of organizations (elementary particles, atoms and molecules, large systems, biological systems) each of which is considered at specific pre-determined external constraints. The most serious drawback of that approach is that the links between levels are broken. Further, it is assumed that the hierarchical constraints go bottom up, i.e. from elementary particles to living organisms. On the contrary, the boundedness appears as principle of organization which allows hierarchical self-organization at non-specified external constraints which is maintained both bottom up and top down by means of loop-like inter-level feedbacks that serves as constraints imposed by one level to the nearest (both upper and lower) ones.

One of the key differences between both approaches is that while the traditional one allows algorithmic reducibility and thus opens the door to search for a Universal Law, the present approach makes any specific law unique and algorithmically irreducible to any other one as a result of the highly non-trivial role of the hierarchical self-organization.

TIME SERIES INVARIANTS AND EARLY WARNING

So far we have considered the most general conditions and characteristics of a stable behavior in the frame of the concept of boundedness. Our ambition is to provide a systematic study of all aspects of that concept. Key point in the introduced concept is the renounce of the highly popular idea of the environmental impact predetermination. Indeed, the traditional approach implements predetermination by means of setting it as a subject of some apriori given probabilistic distribution. A recent development of the probabilistic approach is set on the use of unknown distributions. Nonetheless, the common framework of all probabilistic approaches is that the increments and the size of the successive impacts are left arbitrary; only the frequency of their appearance must follow the apriori set distribution. The latter assumption, better known as the Large Numbers Law, tacitly implies lack of correlations among nearest and any other successive impacts. This is possible if only the result of every impact is a return to the one and the same state called equilibrium. This sets a match between the environmental influences and the fluctuations exerted by a system. However, this way the apriori determined distribution sets both the environment and the response deprived of any identification and individuality: what is expected as a result of any environmental impact

is a universal type of response. This approach commences from the study of the ideal gases for which it turns remarkably correct. Later its validity has been expanded on the grounds of taken for granted arguments about its relevance as one of the basic scientific postulates. Yet, the prevailing argument, though usually only tacitly implied, is that this setting allows relatively easy "predicting" of the future behavior.

On the contrary, the concept of boundedness is grounded on the two basic assumptions: (i) the first one is that the response is local and specific; so it is determined only by the current impact and the current state; (ii) the second one represents a completely novel understanding of the idea about equilibrium: we assume that a system could stay stable at every environmental impact if only the variations of the latter are confined within certain margins. Moreover, we assume that there is a one-to-one correspondence between the current characteristics of the environmental impact and the corresponding stable state. Therefore, it is not necessary for a system to "return" to a single state, equilibrium, after every impact. Yet, our environmental impact and the corresponding response are not let to be completely arbitrary; we impose the condition of boundedness as a necessary condition for the long-term stability. The obvious argument in favor of imposing bounds over the rates of involving and exchanging energy/ matter with the environment is that, if otherwise, a system would develop local defects, strains etc., which in due time would result in its breakdown. It should be stressed that the boundedness over the rates of exchanging and involving matter/ energy constitutes the major distinction between the concept of boundedness and the traditional approach: indeed, the traditional approach lets the succession of impacts free: thus a system turns open at every moment to experience large enough for its destabilization impact. Note that since the succession is arbitrary, this happens in a "catastrophic" way, i.e. without any "signs" of early warning.

But how about the boundedness? On the one hand, taking into account the correlations between successive responses introduced by the boundedness it is highly likely to expect that the development of any change in the behavior is an extended in the time phenomenon. Thus, it is to be expected that it signals out some early warning signs. The questions is whether there exists certain classification of those signs so that to know in advance and with certainty whether any "hazardous" for the stability process develops. One the other hand, the obtained so far results seem to say definitely "NO". Indeed, as already established in the previous Chapter, the key property of the time-series invariant called K-entropy is to put a ban over the predictability of both the nearest any farther future behavior. Yet, this ban is expected and completely comprehensible in view of the release of the local response from any predetermination. Thus, the local characteristics of the response are to be unpredictable. However, we also have proven that the K-entropy of a BIS is always finite on the contrary to a random sequence where it is infinite. What makes the difference is the degree of the unpredictability: while the infinity for a random sequence implies zero probability for any predictability, the finite value of the K-entropy implies a finite value for predictability! Thus, K-entropy acquires a new comprehending: it could be considered as a universal uncertainty for the prediction rather than a ban over the predictability.

The first task in this line of considerations about setting an early warning sign for a hazardous event is how to distinguish between a temporary "deviation" from the steady behavior and the development of any other change in the behavior. Obviously, K-entropy is of no help for this task. Yet, the non-zero certainty of prediction suggests presence of a global component in the response. We already put forward the assumption that the global component of the response is to be associated with self-organized pattern whose major property is to retain its characteristics the same in a specific domain of environmental variations. Because of this property we called the corresponding pattern "homeostasis". And, as we have already established, the homeostasis is characterized by the discrete band in the power spectrum of the corresponding time series. Now, it becomes evident that systematic changes in the behavior signal out a specific line which does not belong either to that of the self-organized pattern associated with the homeostasis or with the continuous noise band. Further, there are three possible scenarios:

1. The line does not persists – then, we can conclude that this is a temporary deviation;
2. The frequency of the line is rational with respect to the frequencies in the discrete band; this implies that a resonance will develop in due time. In general, a resonance is to be associated with accumulating and/or exchanging very large amount of energy/matter and thus it is very likely to expect damages. Then, the appearance of any line with rational frequency is better to be considered as an early warning sign for a hazardous event;
3. The frequency of the line is irrational with respect to the frequencies in the discrete band; then no resonance is expected; the presence of such line is to be considered a sign of a development of a new quality and/or it could be a sign of changes in other hierarchical levels. In both cases the irrationality of the frequency implies that the event is not hazardous. Yet, it appears as a herald of diversity, herald of the "birth" of a new quality, herald of the "birth" of a new law.

CONCLUSION

One of the most intriguing properties of the stable complex systems is the persistent separability of their response to a specific and a universal part. Further, one of the greatest mysteries is that the ubiquity of that separation includes both intelligent and non-intelligent systems. This tempts scientific community to preclude any general reason for explaining the persistency of the above separation. However, we do not share that view: on the contrary we put forward the idea that, though indeed there is no general rule to govern the complex systems behavior, there do exist general *operational protocol* that governs their behavior. Our reason for adopting this point of view is the clear indication that only *stable* complex systems retain common behavior.

We specify the notion of stability by assuming that it is sustainable if and only if both the impact and the response are kept bounded so that neither the rates of exchanging matter/energy with the environment nor the amplitude of these exchanges exceed certain specific for the systems and the process thresholds. Alongside, we release the impact and the response from any pre-determination, i.e. we assume that the response is determined by the current local impact and the current local state only.

This assumption delineates the fundamental difference between our approach and the traditional probabilistic one. To remind, the probabilistic approach sets the environmental impact subject of a pre-determined distribution. Further, according to the linear response theory which is widely applied, the response automatically follows the distribution of the impact. Alongside, in order to provide the stability of the system, it is considered that after each session of impact, the system returns to a single state, called equilibrium. Thus, both the impact and the response are deprived from any individuality and specificity. Then, is it ever possible to expect arising of any intelligent-like behavior in this setting?! On the contrary, the concept of boundedness regards the response as highly specific local characteristics of the system. Moreover, the permanent boundedness opens the door to self-organized collective response which we associate with the specific part of the response. Indeed, the persistency of a self-organized collective response is proven by

means of the exclusive additive decomposition of the power spectrum of a BIS. Indeed, as proven in the present Chapter, the steady pattern of the response of a stable complex system, presented through a BIS, maintains its characteristics with constant accuracy which is insensitive to the statistics of the corresponding environmental impact. Consequently, this result gives not only a credible general explanation for the persistent coexistence of specific and universal properties, but substantiates the use of steady patterns as "letters", i.e. as building blocks of semantics. Note, that this component is not reducible to any weighted average as it would be in the probabilistic approach.

It is worth noting that the differences between the probabilistic approaches and the concept of boundedness are conceptual, i.e. it is not that the concept of boundedness renders the probabilistic approach irrelevant but it is rather that it renders their understanding and utilization to acquire novel comprehending. The novel status of the probabilistic approach will be discussed in section *Invariant Measure. Power Laws* in Chapter 5.

Thus, the separation of the response to a specific and universal part stands as a universal and persistent property of stable complex systems. Yet, the major move ahead is the delineation of the difference between the intelligent and non-intelligent systems in the general setting of the boundedness. We put forward the idea that this discrimination is grounded on the following novel understanding of the notion of information: a piece of information is any certain knowledge which provides unique "identification" of an object from any other in an ever-changing environment. To meet this task we assume to characterize information symbols with the discrete bands of the corresponding power spectra. Indeed, their exclusive generic property to comprise a specific non-recursive component makes the algorithmic unreachablity of any information symbol from any other available. In turn, the algorithmic "uniqueness" of the encapsulated information makes the "identification" of the corresponding object unique. The non-recursive component is brought about by the highly non-trivial interplay between the self-organized component of the response (whose characteristic is the discrete band in the power spectrum) and its "stochastic" counterpart (whose characteristic is the continuous part in the power spectrum).

It is worth noting that, though the information symbols are algorithmically unreachable, the physical ("hardware") implementation of the non-recursive computing happens in finite number of steps and by means of involving/exchanging finite amount of energy and matter. This substantiates the major difference with a Turing machine where hardware silently "follows" the algorithmic properties of the software. Then: (i) the greatest value of a Turing machine is that it can execute any recursive algorithm; however, this comes at the expense of non-autonomous comprehending of the obtained output. On the contrary, the ability of the introduced by us non-recursive computing to execute any algorithm is constrained by the boundedness which precludes transitions which exceed thresholds; at the same time, it substantiates the fundamental value of that approach – autonomous comprehending. Thus, the Turing machine and the non-recursive computing appear rather as counterparts then as opponents.

The major delineation between the non-recursive computing and the traditional information theory is that the proposed by us definition of an information symbol relates its information content with a specific law of Nature, that of the corresponding self-organized pattern characterized by the same discrete part of the power spectrum. A distinctive property of all these "laws" is that each of them admits the principle of relativity which asserts that neither law of Nature depends on any preferred choice of the reference frame. Here, the principle of relativity is met by the proof about the permanent maintenance of constant accuracy

for the reoccurrence of the discrete band. Thus, not only the process does not select and/or signal out any preference time point but this happens in an ever-changing environment, i.e. even though the environment does not re-occur the same, the law stays intact and its characteristics are knowledgeable with the same accuracy. Thus, it is to be expected that the hierarchical self-organization of complex systems would exert not only the consistency of one "law" and another but the creation of new ones. Moreover, we are able to specify the early warning signs for the creation of new laws.

It should be stressed that the presented approach opens the door to radically novel viewpoint on the issue whether the Universe is computable. Our claim is that the algorithmically unreachability of the laws of the Universe from one another automatically discards the view that the Universe is knowledgeable by means of Turing machines-type computing.

Thus, the first step on the road leading from the boundedness to self-organized semantics turns very promising: we have been able not only to explain successfully the persistent coexistence of specific and universal part of the response of the complex systems by means of imposing only the mild constraint of boundedness but to put the grounds for non-recursive computing. Yet, what we still lack is the achievement of an autonomous executing and comprehending of the information encapsulated in it. This constitutes our next major task: to prove that an autonomous comprehending of non-recursive information is available if and only if the non-recursive computing is organized and executed in a semantic-like way. Thus, we have to find out how the other ingredients of a semantic organization: grammar rules, punctuation marks, etc. are related to the properties of complex systems.

Chapter 3
Hierarchical Order I:
Inter–Level Feedback on Quantum Level

ABSTRACT

The purpose of the hierarchical organization is to strengthen the response by means of two general implements: (i) specification of multi-level structure so that each level to respond to specific impacts; (ii) the levels cooperate one with another by means of inter-level feedbacks. The role of the inter-level feedbacks is to sustain the response of any given level bounded by means of keeping local amplifications, local damping, and other non-linear effects restrained. The general purpose of the inter-level feedbacks to sustain long-term stability suggests that they must obey boundedness. Further, the ubiquity of the universal properties of the complex systems promptly suggests that the inter-level feedbacks must appear as a bounded "environment" for every hierarchical level. The non-trivial application of the concept of boundedness to quantum phenomena is considered.

INTRODUCTION

The major goal of the present book is to present a credible systematic theory whose central theme is the answer to the question what makes a complex system intelligent and why it should share certain universal, yet indifferent to the intelligence properties. The first two chapters deal with the second part of the question. It puts forward the assumption that the universal properties are to be straightforwardly associated with long-term stability of a system. Thus, it has been proven that the concept of boundedness alone, viewed as the necessary condition for long-term stability, provides the existence of 3 time series invariants as a generic property of every zero-mean bounded

DOI: 10.4018/978-1-4666-2202-9.ch003

irregular sequence (BIS). A distinctive property of these time series invariants is that each of them is insensitive to the environmental statistics and to the operation of coarse-graining. This implies that the obtained knowledge about the behavior of a complex system is independent from the lack of a global pre-determination of the response and the inevitable distortions in its recording.

From the above considerations it becomes clear that for the first time we encounter the highly non-trivial interplay between the issue about boundedness viewed as a condition providing long-term stability, and the issue about information that generates its most general meaning. By following these considerations further, we again encounter that interplay when discussing the separability of a power spectrum to a discrete and a continuous part. Indeed, by proving constant accuracy of the discrete band reoccurrence in an ever-changing environment, we actually prove that the information encapsulated in it is reproducible with well defined certainty. Yet, an additional reading of that reproducibility is that no information can be created by variations (noise) only. This conclusion is strongly supported by the universal finiteness of another already defined time-series invariants, K-entropy, which comes to say that the response is a two-component variable: the first component is the above discussed noise one whose future behavior is unpredictable; and the second component is a collective self-organized pattern which remains steady and robust under the "noise" part. Thus, we can conclude that the only component which carries information is the steady one. Further, by associating the information with the discrete band in the power spectrum, the constant accuracy of its reoccurrence retains the following two-fold meaning: (i) the information cannot be created or destroyed by the noise; (ii) the energy/matter involved in its creation and sustaining is constant which is always less than the all energy/matter involved in the overall response. As a matter of fact, the overall energy/matter involved and exchanged by the response is given by the overall

power spectrum; and since the latter comprises additively two components, a discrete one and a noise one, it is obvious that the energy involved in the discrete one, i.e. one associated with the information, is always less than 100% .

Next in this line of reasoning we found out that the minimal necessary condition for the noise band in the power spectrum so that it to posses the property not to signal out any specific component is the presence of the same metrics for all components. This requirement goes consistently with our suggestion that the response is determined only by the current local impact and the current state of a system. A minimal condition for sharing the same metrics by components in the discrete and the continuous band is to be described by the same variable. The next important element in this argumentation is to put forward the assumption that the interplay between both parts is substantiated by the presence of a non-recursive line in the discrete band. The importance of the latter is enormous since it fundamentally changes the entire view on the relation between different phenomena and laws in Nature. The presence of a non-recursive component in the discrete band viewed as a generic property implies algorithmic unreachability of one law, viewed as a rule for organizing and sustaining of a certain self-organized pattern. Put it in other words, each law becomes algorithmically uncompressible. In turn, as discussed in the previous Chapter, this implies renounce of the idea about existence of Universal law to which otherwise the idea about algorithmic compressibility in the traditional information theory leads.

Yet, so far our theory seems rather axiomatic and thus requires development in several directions: to find out the role of hierarchical structuring in the interplay between the issue about long-term stability and information in its widest sense; to find out the energy costs of the information and eventually to answer the major question whether the obtained intelligence has autonomous comprehending and if so, is it the boundedness that guarantees it. The first step in substantiating these axiomatic statements is to expand the general idea about the interplay between the boundedness viewed as the necessary condition for long-term stability and the information in its widest meaning; for this purpose we should derive a systematic view on the role of hierarchical structuring in that interplay.

BACKGROUND

It appears that we start our journey with a problem: on the one hand, we require locality for the response and thus it is naturally to expect its description by the same variable for the noise component and for the "homeostatic" one; on the other hand, we expect radically different behavior for the noise component and its homeostatic counterpart: while the noise component is highly sensitive to the environmental variations, the distinctive property of the homeostatic one is robustness to the environmental variations. The reconciliation of these two opposing assumptions comes from a new look on the problem viewed from the following perspective: let us consider the above problem together with the question what renders the environmental impact bounded. Our answer to it is grounded on the non-trivial role played by the hierarchical organization. We assert that its purpose is to strengthen the response by means of two general implements: (i) specification of multi-level structure so that each level to respond to specific impacts; (ii) the levels cooperate one with another by means of inter-level feedbacks. The role of the inter-level feedbacks is to sustain the response of any given level bounded by means of keeping local amplifications, local damping and other non-linear effects restrained. The general purpose of the inter-level feedbacks to sustain long-term stability suggests that they must obey boundedness. Further, the ubiquity of the universal properties of the complex systems promptly suggests that the inter-level feedbacks must appear as a bounded "environment" for

every hierarchical level. Then, shall we expect separability of the hierarchical levels from one another?! This is highly non-trivial question and the answer will be that for the intelligent-like systems it is negative. Yet, the affiliation of both intelligent and non-intelligent systems to the same class of complex systems suggests classification of the inter-level feedbacks into two classes: those which are common for both intelligent and non-intelligent systems and those which are specific for the intelligent ones only. The task of this Chapter is to consider the first ones while the specific for the intelligent systems feedbacks will be considered in Chapter 7.

An immediate outcome of the above presented perspective is that the specification of the impacts according to the hierarchy of matter organization suggests corresponding hierarchy of execution and functional organization of the information associated with it. The first step in this correspondence is that the specification of the impacts imposes specific enumeration of the control parameter space where the control parameters enumerate the environmental impact in a certain order. As a consequence, this enumeration sets one-to-one correspondence between the state space of the corresponding system and its control parameter space. This mapping is implemented by specific complex non-linear interplay among the internal processes of the system, environmental impact and inter-level feedbacks. The presence of this mapping constitutes the fundamental difference from the traditional statistical mechanics where the properties of any macroscopic state is characterized by a probability for its occurrence alone. The latter would be possible only in a state space where the "jumps" between successive states are not restrained; then the state space would neither need nor would specify any metrics and in consequence the only characteristic is a single number which is assigned so that the Large Numbers Law to hold. On the contrary, the enumeration of the control parameter space pre-supposes metrics which defines not only the position of a state but the distance to any other. The requirement about metrics is met by the boundedness as we already discussed it in the previous Chapter. The fact is the constant accuracy of discrete band reoccurrence in the power spectrum holds if and only if the response retains a unified metrics.

Next we need to consider the question how the assumed by us hierarchical specification of the response in the sense that it is distributed among hierarchical levels so that each level to respond only to specific for it impacts turns itself is not the main implement for a non-extensive reduction of the degrees of freedom. Indeed, an obvious condition for substantiation of the impact distribution among the hierarchical levels is the "partitioning" of the impacts so that a domain of control parameters at a given level to corresponds to a point on the next. The distinctive property of that partitioning is that it happens in the enumerated space of control parameters. It turns out that on one the hand, each domain must be characterized by a specific intra-domain invariant, i.e. a characteristic that remains the same inside the entire domain; on the other hand, the location of the domain in the control parameter space is metrically specified. Then our intuition persuades us to suggest association of intra-domain invariants with information symbols. In order to substantiate this suggestion it is necessary to answer a lot of questions among which the most important are: are the intra-domain invariants characterized by discrete bands of the corresponding power spectra? Do they comprise a non-recursive component and does it indeed come from an inter-level feedback? And if so, what makes these properties generic for the hierarchy of the response?

We already have discussed the issue that the steady component of the response is to be associated with a self-organized pattern. The idea of self-organization is one of the most brilliant achievements of the modern interdisciplinary science. It is very useful for us because it meets some of our expectations: it defines enumeration of the control parameter space and sets domains characterized by specific invariants that are the

discrete bands in the corresponding power spectra. However, the theory is incomplete because the traditional approach to it is vulnerable to the account for the variability of the environment and/ or presence of noise. The major flaws are that by taking into account the noise the approach becomes non-generic and does not automatically provide constant accuracy of the discrete band reoccurrence. In turn, this makes the corresponding pattern unable to meet the principle of relativity.

The merits and the setbacks of the traditional theory of self-organization constitute the task of the present Chapter where we are going to incorporate the idea of boundedness in the theory and to demonstrate that it alone is able to eliminate the setbacks and to yield a generic approach which retains all desired properties. In the next Chapter it will be proven that the boundedness set on a given hierarchical level, spontaneously emerges at the next.

Since both intelligent and non-intelligent systems are made of atoms and molecules we start our consideration with that level of matter organization. The question is how atoms and molecules are to be organized so that to "define" macroscopic variables such as concentrations and temperature. Our discussion will be focused mainly on the issue how and why boundedness is to be incorporated in the current theory of self-organization. The task of the next section is to make evident that the root of the problem lies in the dynamics. The next task is to consider in details how the boundedness incorporated in the dynamics gives rise to an inter-level feedback.

ROOTS OF THE PROBLEM

Amplification of the Local Fluctuations

The self-organization is one of the most advanced and fascinating hypothesis about the behavior of the extended many-body systems with short-range interactions. It asserts that the emergent properties on macroscopic scale, obtained as a result of self-organization, are insensitive to the microscopic dynamics and are independent of the individual elements and components. The mild conditions of the derivation of the reaction-diffusion equations (hyperbolic partial differential equations) that serves as mathematical tool for its description support the ubiquity of its application: the idea of self-organization is applied to wide spectrum of systems as diverse as chemical systems, optical lattices, social systems, population dynamics etc. Indeed, these equations are founded on the postulate that the diffusion does not create or annihilate substance which provides the conservation of substance flow in space-time. However, the rigorous derivation of the reaction-diffusion equations requires explicit demonstration that the conditions at which the microscopic dynamics does not contribute to the emergent properties are as credible as the idea of flux conservation. In particular, one such condition is that the local fluctuations are to be automatically damped. Moreover, it requires that damping happens at time scale much smaller than the specific scale on which the state variables change significantly. This is a crucial point, since the separation of the time scales to fast and slow ones provides spatio-temporal continuality of the extensive state variables such as concentrations, temperature etc. so that the flux of substance to be expressed in terms of those variables. In turn, this ensures the target independence of the emergent properties from the microscopic dynamics. But is the separation of the time scales grounded on the same plausible basis as the substance conservation? Our first aim is to demonstrate that, to the most surprise, it is not. Next we provide evidence that the basic models that are in circulation so far approach the separation of the time scales at the expanse of letting the velocity of transmitting substance trough space to be arbitrary. Note, that arbitrary means that the velocity could be greater than the speed of light! On the other hand, as we will demonstrate further, the boundedness of velocity alone is not

sufficient to make the self-organization available because it participates in exponential amplification of the local fluctuations. That amplification is a novel effect, considered for the first time in the present section, that is a result of the interplay between the boundedness of the velocity and the lack of correlations among local fluctuations due to short-rang interactions. In turn, the amplification of the local fluctuations assisted by the lack of correlation among them results in triggering of local destabilization that if not eliminated, would yield system breakdown. Thus our aim is to prove that the concept of boundedness is not reducible to boundedness of the velocities alone.

We shall illustrate that short-range interactions give rise to permanent destabilization that is locally amplified even when the velocity ansatz holds. To elucidate better the problem let us first make apparent how the self-organization appears when the velocity of transmitting substance is let to be arbitrary. For this purpose let us follow (Gardiner, 1985) in the derivation of the diffusion equation:

Let $f(x,t)$ be the number of entities per unit volume. We compute the distribution of entities at the time $t+\tau$ from the distribution at time t. Let the function $\varphi(\Delta)$ be the frequency of the jumps to distance Δ back and forth a given point. Then, the flux conservation requires that the number of entities which at time $t+\tau$ are found between two planes perpendicular to the $x-axis$ and passing through points x and $x+dx$. One obtains:

$$f(x,t+\tau)dx = dx\int_{-\infty}^{+\infty} f(x+\Delta)\varphi(\Delta)d\Delta$$

(3.1)

Under the supposition that τ is very small, we can set:

$$f\left(x, t + \tau\right) = f\left(x, t\right) + \tau \frac{\partial f}{\partial t} \qquad (3.2)$$

Furthermore, we develop $f\left(x + \Delta, t\right)$ in powers of Δ :

$$f\left(x + \Delta, t\right) = $$
$$f\left(x, t\right) + \Delta \frac{\partial f\left(x, t\right)}{\partial x} + \frac{\Delta^2}{2!} \frac{\partial^2 \left(x, t\right)}{\partial x^2} + \ldots$$
$$(3.3)$$

We can use this series under the integral:

$$f + \frac{\partial f}{\partial t} \tau = $$
$$f \int_{-\infty}^{+\infty} \varphi\left(\Delta\right) d\Delta + $$
$$\frac{\partial f\left(x, t\right)}{\partial x} \int_{-\infty}^{+\infty} \Delta \varphi\left(\Delta\right) d\Delta + \qquad (3.4)$$
$$\frac{\partial^2 f\left(x, t\right)}{\partial x^2} \int_{-\infty}^{+\infty} \frac{\Delta^2}{2} \varphi\left(\Delta\right) d\Delta + \ldots$$

Because $\varphi\left(x\right) = \varphi\left(-x\right)$, the second, fourth etc. terms on the right-hand side vanish, while out of the 1st, 3rd, 5th etc., terms, each small compared with the previous. We obtain from this equation, by taking into account consideration:

$$\int_{-\infty}^{+\infty} \varphi\left(\Delta\right) d\Delta = 1 \qquad (3.5)$$

and setting

$$\frac{1}{\tau} \int_{-\infty}^{+\infty} \frac{\Delta^2}{2} \varphi\left(\Delta\right) d\Delta = D \qquad (3.6)$$

Keeping only the 1st and third term of the right-hand side,

$$\frac{\partial f}{\partial t} = D \frac{\partial^2 f}{\partial x^2} \qquad (3.7)$$

In result, the microscopic derivation yields the flux conservation. Moreover, the dependence on the microscopic dynamics is encapsulated in a single parameter D that comprises asymptotic statistical property, namely the variance of the jump length. However, the circle is not yet closed because the above derivation comprises additional supposition that has not been considered. In fact let us have a closer look on the integration in (3.1) and (3.4). The philosophy of (3.4) asserts that in a small but otherwise arbitrary time interval τ the flux that crosses the boundary between x and $x + dx$ is the average between jumps back and forth the layer. Yet, the range of integration of the jump length $\left(-\infty, +\infty\right)$ implicates independence of the jump length Δ and the duration of the time interval τ from one another. In turn, that independence renders the microscopic dynamics to be averaged out on times scales larger than certain one. However, this brings about an apparent contradiction: on the one hand, the credibility of the expansions in (3.1)-(3.4) implies existence of certain time scale t_{fast} over which the microscopic dynamics is averaged out. Therefore, the value of τ is to be set larger than t_{fast}. On the other hand, the notion of the diffusion coefficient D in (3.7) requires τ to be smaller than t_{fast}. Moreover, the existence of D is possible if and only if Δ and τ are related; otherwise, the lack of relation between Δ and τ renders D τ — dependent provided the variance of the jump length exists. Thus, one and the same τ participates to the same equation in two confronting each other roles.

In addition, the independence of the length of a jump Δ and its duration τ suffers severe physical disadvantage: it renders the velocity $\frac{\Delta}{\tau}$ of transmitting substance trough space to be arbitrary; in particular it can become even greater than the speed of the light!

Note, that the lack of relation between Δ and τ is plausible only under the condition of perfect stirring which asserts that stirring is so intensive that every entity can be found with equal probability anywhere in the system. However, the stirring is not available for a number of systems such as interfaces, optical lattices, social systems, population dynamics etc. The other interpretation of the perfect stirring, namely that the system has already arrived at local thermodynamical equilibrium, does not help reconciliation because it does *not* point out the route through which the microscopic dynamics ensures the thermodynamical equilibrium.

So, in order to put the hypothesis about self-organization on stable ground, the starting point must be the postulate that whatever the system is the velocity of transmitting substance and/or energy is to be bounded (velocity ansatz).

The question that immediately arises is whether the requirement about relation between Δ and τ, i.e. velocity ansatz, is enough to save the idea of self-organization. The goal of the present section is to demonstrate that: (i) the dependence between Δ and produce local fluctuations; (ii) due to short-range interactions, the lack of correlations among them makes possible their amplification. In result, the latter not only violates the idea of self-organization but gives rise to permanent destabilization of the system that if not eliminated, rapidly yields its breakdown. In the next section we will demonstrate that the elimination of the destabilization is possible only under radically novel viewpoint on a number of issues.

Let us first see how the diffusion equation is modified under the postulate about boundedness of the velocity. Obviously, it makes Δ function

of $\tau : \Delta = \Delta(\tau)$. Then, (3.1) is modified as follows:

$$f(x, t+\tau)dx = dx \int_{-\varsigma(\tau)}^{+\varsigma(\tau)} f(x + \Delta(\tau))\varphi(\Delta(\tau))d\tau$$

(3.8)

where $\varsigma(\tau)$ is the maximum possible jump length at given τ. Accordingly, Equation (3.4) becomes:

$$f + \frac{\partial f}{\partial t}\tau =$$
$$f \int_{-\varsigma(\tau)}^{+\varsigma(\tau)} \varphi(\Delta)d\Delta + \frac{\partial f(x,t)}{\partial x} \int_{-\varsigma(\tau)}^{+\varsigma(\tau)} \Delta(\tau)\varphi(\Delta(\tau))d\tau +$$
$$\frac{\partial^2 f(x,t)}{\partial x^2} \int_{-\varsigma(\tau)}^{+\varsigma(\tau)} \frac{\Delta^2(\tau)}{2}\varphi(\Delta(\tau))d\tau + \dots$$

(3.9)

The fundamental difference between (3.4) and (3.9) is that the averaging in the right-hand side of the integrals in (3.4) specifies only the range of the jump length leaving the time interval arbitrary while in (3.9) the averaging explicitly involves the dependence between the jump length and its duration. In result, the averaging over a finite time interval yields:

$$\int_{-\varsigma(\tau)}^{+\varsigma(\tau)} \Delta(\tau)\varphi(\Delta(\tau))d\tau = \beta(x, t, \varsigma(\tau)) \quad (3.10)$$

where $\beta(x, t, \varsigma(\tau))$ is non-zero and changes its sign from positive to negative depending on the space-time location of the window $\varsigma(\tau)$. To compare, the averaging over infinite interval makes $\beta(x, t, \varsigma(\tau)) \equiv 0$ at every spatio-temporal point (x, t):

$$\beta\left(x,t,\zeta\left(\tau\right)\right) = \int\limits_{-\infty}^{+\infty} \Delta\varphi\left(\Delta\right) d\Delta \equiv 0 \qquad (3.11)$$

Accordingly:

$$\int\limits_{-\zeta(\tau)}^{+\zeta(\tau)} \frac{\Delta^2\left(\tau\right)}{2} \varphi\left(\Delta\left(\tau\right)\right) d\tau = D\left(x,t,\zeta\left(\tau\right)\right) \qquad (3.12)$$

where the value of $D\left(x,t,\zeta\left(\tau\right)\right)$ varies with the size and location of the windows. In result, the equation for $f\left(x,t\right)$ becomes:

$$\frac{\partial f\left(x,t,\zeta\left(\tau\right)\right)}{\partial t} =$$
$$\beta\left(x,t,\zeta\left(\tau\right)\right)\frac{\partial f\left(x,t,\zeta\left(\tau\right)\right)}{\partial x} + \qquad (3.13)$$
$$D\left(x,t,\zeta\left(\tau\right)\right)\frac{\partial^2 f\left(x,l,\zeta\left(\prime\right)\right)}{\partial x^2}$$

The fundamental difference between (3.7) and (3.13) is that the window parameters ζ and τ in (3.13) are entangled with the spatio-temporal variables x and t while in (3.7) the window parameters do not participate at all. However, the entanglement of the window parameters and the spatio-temporal variables strongly interferes with the idea that the emergent properties on macroscopic scale are to be independent of the corresponding microscopic dynamics. Besides, it opens the door to the following option: since the lack of correlations among local fluctuations renders $\beta\left(x,t,\zeta\left(\tau\right)\right)$ and $D\left(x,t,\zeta\left(\tau\right)\right)$ to be irregular functions, is it possible that their stochasticity brings about amplification of the local fluctuations that in turn triggers local destabilization followed by rapidly developed global one. In result we face a fundamental problem: on the one hand, in order to save the idea of self-organization we must save the supposition about the existence of two time scales t_{slow} and t_{fast} so that:

$$f\left(x,t,\zeta\left(\tau\right)\right) = f_{slow}\left(x,t\right) + f_{fast}\left(x,t,\zeta\left(\tau\right)\right)$$
$$(3.14)$$

(ii) on the other hand, the supposition about separability of the times scales is justified only by automatic damping of the local fluctuations. Next, we will demonstrate that the fulfillment of both requirements cannot be provided by the velocity anzatz alone. We proceed through the opposite, namely we suppose *apriori* that the separation of the time scales holds. Our task is to prove that it does not prevent amplification of the local fluctuations.

The separation of the time scales implies that for the time scales Δt such that $t_{slow} >> \Delta t >> t_{fast}$, the following averaging yields:

$$\int\limits_{0}^{\Delta t} f_{fast}\left(x,t,\zeta\left(\tau\right)\right) d\tau = 0$$
$$\int\limits_{0}^{\Delta t} \frac{\partial f_{fast}\left(x,t,\zeta\left(\tau\right)\right)}{\partial x} d\tau = 0 \qquad (3.15)$$
$$\int\limits_{0}^{\Delta t} \frac{\partial^2 f_{fast}\left(x,t,\zeta\left(\tau\right)\right)}{\partial x^2} d\tau = 0$$

The condition (3.15) actually imposes uniform convergence of $f_{fast}\left(x,t,\zeta\left(\tau\right)\right)$ and its first and second derivative to $f_{slow}\left(x,t\right)$ and its corresponding derivatives. Note that (3.15) is justified whenever the local fluctuations are damped on the time scales smaller than t_{fast} : if so, the distance between $f_{fast}\left(x,t,\zeta\left(\tau\right)\right)$ and $f_{slow}\left(x,t\right)$ is always finite which in turn verifies both the separation in (3.14) and the uniform convergence set by (3.15). As an immediate consequence of (3.14) and (3.15), the dependence of $f\left(x,t,\zeta\left(\tau\right)\right)$ on the microscopic dynamics, i.e. on the window parameters $\left(\zeta,\tau\right)$ and their relation, is limited to the time intervals smaller than t_{fast} . Now we come to the question at what conditions on $\beta\left(x,t,\varsigma\left(\tau\right)\right)$ and $D\left(x,t,\varsigma\left(\tau\right)\right)$ (3.14)-(3.15) hold. Let us now substitute $f\left(x,t,\varsigma\left(\tau\right)\right)$ from (3.14) in (3.13). It is obvious that a necessary condition for (3.15) to hold is that $\beta\left(x,t,\varsigma\left(\tau\right)\right)$ and $D\left(x,t,\varsigma\left(\tau\right)\right)$ additively decompose to slow and fast parts:

$$\beta\left(x,t,\varsigma\left(\tau\right)\right) = \beta_{slow}\left(x,t\right) + \beta_{fast}\left(x,t,\varsigma\left(\tau\right)\right) \tag{3.16}$$

$$D\left(x,t,\varsigma\left(\tau\right)\right) = D_{slow}\left(x,t\right) + D_{fast}\left(x,t,\varsigma\left(\tau\right)\right) \tag{3.17}$$

In general, both $\beta_{fast}\left(x,t,\varsigma\left(\tau\right)\right)$ and $D_{fast}\left(x,t,\varsigma\left(\tau\right)\right)$ must be irregular functions that fulfill the following condition: their average over window of any length greater than t_{fast} uniformly converges to $\beta_{slow}\left(x,t\right)$ and $D_{slow}\left(x,t\right)$ correspondingly. This requirement opens the door for $D_{fast}\left(x,t,\varsigma\left(\tau\right)\right)$ to change sign. We will demonstrate that the negative values of the diffusion coefficient implement the amplification of the local fluctuations. Furthermore, the condition that the walk is symmetric renders $\beta_{slow}\left(x,t\right) \equiv 0$. Then, Equation (3.13) becomes:

$$\frac{\partial f_{slow}\left(x,t\right)}{\partial t} + \frac{\partial f_{fast}\left(x,t,\varsigma\left(\tau\right)\right)}{\partial t} =$$
$$\beta_{fast}\left(x,t,\varsigma\left(\tau\right)\right)\frac{\partial f_{slow}\left(x,t\right)}{\partial x} +$$
$$\beta_{fast}\left(x,t,\varsigma\left(\tau\right)\right)\frac{\partial f_{fast}\left(x,t,\varsigma\left(\tau\right)\right)}{\partial x} +$$
$$+\begin{pmatrix} D_{slow}\left(x,t\right) + \\ D_{fast}\left(x,t,\varsigma\left(\tau\right)\right) \end{pmatrix}\begin{pmatrix} \frac{\partial^2 f_{slow}\left(x,t\right)}{\partial x^2} + \frac{\partial^2 f_{fast}\left(x,t,\varsigma\left(\tau\right)\right)}{\partial x^2} \end{pmatrix} \tag{3.18}$$

Let us now average over an interval Δt such that $t_{slow} >> \Delta t >> t_{fast}$. Then, on the condition

for uniform convergence, (3.18) decouples as follows:

$$\frac{\partial f_{slow}\left(x,t\right)}{\partial t} = D_{slow}\left(x,t\right)\frac{\partial^2 f\left(x,t\right)}{\partial x^2} \tag{3.19}$$

and

$$\frac{\partial f_{fast}\left(x,t,\varsigma\left(\tau\right)\right)}{\partial t} =$$
$$\beta_{fast}\left(x,t,\varsigma\left(\tau\right)\right)\frac{\partial f_{fast}\left(x,t,\varsigma\left(\tau\right)\right)}{\partial x} +$$
$$D_{fast}\left(x,t,\varsigma\left(\tau\right)\right)\frac{\partial^2 f_{fast}\left(x,t,\varsigma\left(\tau\right)\right)}{\partial x^2} \tag{3.20}$$

The terms that contains one slow and one fast multiplier becomes zero after averaging over Δt because the corresponding slow term remains constant over that time interval $\Delta t << t_{slow}$. For example:

$$\left\langle D_{slow}\left(x,t\right)\frac{\partial^2 f_{fast}\left(x,t,\varsigma\left(\tau\right)\right)}{\partial x^2} \right\rangle =$$
$$D_{slow}\left(x,t\right)\left\langle \frac{\partial^2 f_{fast}\left(x,t,\varsigma\left(\tau\right)\right)}{\partial x^2} \right\rangle = 0 \tag{3.21}$$

However, the averaging in (3.20) commutes with the differentiation both in time and space only if $f_{fast}\left(x,t,\varsigma\left(\tau\right)\right)$ is always finite; the latter is provided only by automatic damping of the local fluctuations. On the contrary, the alternative of amplification opens the door to unlimited increasing of $f_{fast}\left(x,t,\varsigma\left(\tau\right)\right)$. In order to study that option we have to examine the properties of (3.20) over time scales smaller than t_{fast}. For this purpose let us now present (3.20) in the following slightly modified form:

$$\frac{\partial f_{fast}\left(x,t,\zeta\left(\tau\right)\right)}{\partial t} =$$
$$\frac{\partial}{\partial x}\left[\tilde{D}_{fast}\left(x,t,\zeta\left(\tau\right)\right)\frac{\partial f_{fast}\left(x,t,\zeta\left(\tau\right)\right)}{\partial x}\right] \quad (3.22)$$

where $\tilde{D}_{fast}\left(x,t,\zeta\left(\tau\right)\right)$ is such that:

$$\frac{\partial \tilde{D}_{fast}\left(x,t,\zeta\left(\tau\right)\right)}{\partial x} = \beta\left(x,t,\zeta\left(\tau\right)\right) \quad (3.23)$$

Since both $D_{fast}\left(x,t,\zeta\left(\tau\right)\right)$ and $\beta\left(x,t,\zeta\left(\tau\right)\right)$ are irregular functions with identical statistical properties, $\tilde{D}_{fast}\left(x,t,\zeta\left(\tau\right)\right)$ is again an irregular function that share the same statistical properties.

Actually, the value and sign of $\tilde{D}_{fast}\left(x,t,\zeta\left(\tau\right)\right)$ depends mainly on the window parameters ζ and τ; the dependence on spatio-temporal variables $\left(x,t\right)$ serves to point out explicitly the lack of correlations among their values at closest spatio-temporal locations. This lets us to regard $\tilde{D}_{fast}\left(x,t,\zeta\left(\tau\right)\right)$ as function of the window parameters only. In turn, this consideration let us approximate (3.22) as follows:

$$\frac{\partial f_{fast}\left(x,t,\zeta\left(\tau\right)\right)}{\partial t} =$$
$$\tilde{D}_{fast}\left(x,t,\zeta\left(\tau\right)\right)\frac{\partial^2 f_{fast}\left(x,t,\zeta\left(\tau\right)\right)}{\partial x^2} \quad (3.24)$$

Note that $\tilde{D}_{fast}\left(x,t,\zeta\left(\tau\right)\right)$ is explicitly grounded on the velocity ansatz!

The consideration that $\tilde{D}_{fast}\left(x,t,\zeta\left(\tau\right)\right)$ is set constant equal to its current value opens the door for explicit establishing the solution. The velocity ansatz bounds us to look for the solution in the form of a single variable λ that couples x

and t so that the velocity to be bounded. Then $\lambda = \frac{x}{t^{\alpha\left(x/t\right)}}$ where we set $\alpha\left(\frac{x}{t}\right)$ function because the diffusion coefficient $\tilde{D}_{fast}\left(x,t,\zeta\left(\tau\right)\right)$ varies in the space-time. Therefore, the major term in the solution of (3.24) reads:

$$f_{fast}\left(x,t,\zeta\left(\tau\right)\right) \propto \exp\left(-\frac{x^2}{\tilde{D}\left(x,t,\zeta\left(\tau\right)\right)t}\right) \quad (3.25)$$

Since the value and sign of $\tilde{D}_{fast}\left(x,t,\zeta\left(\tau\right)\right)$ varies form on location to the next, $f_{fast}\left(x,t,\zeta\left(\tau\right)\right)$ is also an irregular function that grows up exponentially when $\tilde{D}_{fast}\left(x,t,\zeta\left(\tau\right)\right)$ is negative. Thus, the corresponding local fluctuations are amplified. Moreover, though the rate of the obtained amplification depends only on the local window parameters, it can happen at every instant and everywhere in the system. In turn, it gives rise to and sustains permanent destabilization of the system.

It should be stressed that the reaction-diffusion coupling is not able to prevent the destabilization since the reaction interactions are local events the rate of which is proportional to the local concentrations of the reactants. So, the local fluctuations are straightforwardly mapped onto the reaction rates. This consideration is easy to be traced in the formal solution of the corresponding reaction-diffusion equations that looks like as follows:

$$\frac{\partial f_{fast}\left(x,t,\zeta\left(\tau\right)\right)}{\partial t} =$$
$$\kappa A\left(f_{fast}\left(x,t,\zeta\left(\tau\right)\right)\right) - \gamma R\left(f_{fast}\left(x,t,\zeta\left(\tau\right)\right)\right) +$$
$$\tilde{D}_{fast}\left(x,t,\zeta\left(\tau\right)\right)\frac{\partial^2 f_{fast}\left(x,t,\zeta\left(\tau\right)\right)}{\partial x^2}$$
$$(3.26)$$

where κ and γ are the control parameters; $A\left(f_{fast}\left(x, t, \zeta\left(\tau\right)\right)\right)$ is the rate by which the substance enters the system ; $R\left(f_{fast}\left(x, t, \zeta\left(\tau\right)\right)\right)$ is the reaction rate. Since both (3.24) and (3.26) are hyperbolic partial differential equations that share the same Green function whose major term is (3.25), the reaction-diffusion coupling indeed does not help preventing the amplification of the local fluctuations.

Outlining, the conclusion drawn from the above considerations is that the velocity anzatz alone, i.e. boundedness over the velocities of the atoms and molecules, is unable to provide automatic damping of local instabilities; moreover, to the most surprise, it yields their amplification! Thus, even though necessary, the velocity anzatz applied in the above axiomatic way is not only inefficient but it turns out even hazardous for the system stability. This result suggests that we should look for additional assumptions so that to achieve the goal.

Necessity of Insensitivity to Partitioning

The growing interest in the traditional approach of self-organization has been provoked by its recent application to informational, social and ecological systems. Despite its considerable potential for explanation of various phenomena, the diversity of dynamical rules strongly challenges its fundamental postulate inferring separation of the time scales such that to render the insensitivity of the macroscopic behavior to the particularities of the dynamics of the constituents. The assumed insensitivity makes possible the association of time-independent frequencies with the development of every macroscopic fluctuation. Therefore the time average on every macroscopic time scale becomes uniquely determined and time-translational invariant. The importance of the latter property is that it provides the reproducibility of the events,

both deterministic and stochastic. If considered as broken time-translational invariance, the manifestation of each event would ambiguously depend on the moment of its realization even when the conditions for its performance remain unchanged. This type of reasoning is in conflict with the reproducibility viewed as the simplest form of causality and by opposing straightforward cause-effect relationships it conditions the crucial importance of establishing the circumstances which validate the aforementioned postulate. As a primary concern for finding solution of this disparity, we will demonstrate that the general condition for its validation confronts both the classical and the quantum dynamics.

In order to outline this problem we will start examining the general prerequisite of the target invariance: the extensive macroscopic variables (e.g. concentration, temperature) are to be invariant with respect to the partitioning of the system; at the same time the intensive macroscopic variables must be additive regardless to the way the partitioning is made. It is beyond doubt that the independence from the partitioning justifies the uniqueness of the averaged product and its independence from the details of the dynamics of the constituents. If otherwise, i.e. in the case of any dependence on the partitioning, the macroscopic variables depend on the choice of the auxiliary reference frame selected to define the concrete way of partitioning and in result this dependence yields violation of the principle of relativity. So far it has been taken for granted that the avoidance of this problem is justified by the assumption that since the macroscopic variables vary much slowly than the fast dynamical ones, the averaging over the entire phase space (ensembles in the classical case) would take place at every location in the system considered. Therefore the same averaged probability is assigned to every constituent. The elusive point, however, resides in the lack of any comprehensive argument whether the averaging protocol uniformly converges to an invariant for the spatio-temporal location value. The importance of this issue lies in the fact that it acts as a criterion that discriminates two fundamentally different conducts: whilst any non-uniform convergence makes the macroscopic behavior critically dependent on the choice of the auxiliary reference frame and thus provokes conflict with time-translational invariance (principle of relativity), the uniform convergence implements elimination of any such dependence. This subject is discussed next where key arguments for the non-uniformity of that convergence are proposed. Moreover, we will prove that the core of the problem is deeply rooted into the dynamics.

The generation of a conflict between the traditional approach to self-organization and the time-translational invariance (principle of relativity) crucially depends on the answer to the question whether the uniform convergence is a generic property of the partitioning. The aim of the present section is to prove that there is no uniform convergence and to demonstrate that the problem is to be traced down to the dynamics. In order to make this proof more evident, we will apply our argumentations to the process of relaxation of adsorbed/absorbed species.

The adsorption/absorption is process such that species confined by the surface/volume relax to the ground state of potential wells situated at certain locations on a surface/volume. The transitions through which the relaxation proceeds, dissipate trough excitation of those collective surface/volume modes which match certain specific for the transition and the mode conditions called hereafter resonance conditions. The excited collective modes, in return, act as a perturbation on the potential well and thus modify the resonant conditions. Since these perturbations are

maintained by the current spatio-temporal configuration, the question is whether they drive any dependence of the probability for relaxation on its morphology. To find out, let us have a look on the energy of the excited modes in an arbitrary location and at arbitrary instant (Box 1):where ω is the frequency of an excited mode; $A(\omega)$ - its amplitude; $\vec{k}(\omega)$ is the dispersion relation of the mode; $\vec{\mu}_i$ is the location of the $i-$th potential well; $\tau_i(x,t)$ is the delay of the mode that travels from the $i-$th potential well.

The procedure that examines the convergence of the partitioning protocol is well established and involves the following steps: (i) considering an window of arbitrary size (ς,θ) located at the point (x,t), the specification of the margins of integration over the frequencies of the excited modes ω and the delays τ that fit the window considered is to be set; (ii) the second step is establishing whether the averaged over that window dissipated energy converges to an invariant for the system value. The intrigue in this procedure is the revelation of a so far overlooked relation between the frequencies and the delays. Indeed, as explained above, any change of the dissipated energy is carried out by those transitions that match the specific for the dynamical process resonance conditions which provide correspondence with the excited collective modes. On the other hand, the excited modes perturb the potential wells and thus modify the resonance parameters. Bearing in mind the general rule of the dynamics that the resonance conditions are highly specific to the process considered the independence of the delays from one another makes

Box 1.

$$J_{exc}(x,t) \approx \sum_i \int_\omega \int_\tau d\omega d\tau A(\omega) \exp\left(i\omega\left(t - \tau_i(x,t)\right) - i\vec{k}(\omega)\bullet\left(\vec{x} - \vec{\mu}_i\right)\right) \tag{3.27}$$

the dissipated energy fine-tuned to the spatio-temporal location which in turn provokes high sensitivity of the averaged product to the window parameters. That is why the averaging drives amplification of even infinitesimal differences in the dissipated energy and so intensifies its non-uniformity. Moreover, the joint action of the fine-tuning and the produced by it non-uniformity brings about additional "scattering" of the delays which results in further escalation of the non-uniformity. Hence, as follows from (3.27), the explicit dependence of the averaged product on the window parameters (ς, θ) and its location (x, t) illuminates the assertion that the non-uniformity is straightforwardly related to the choice of the auxiliary reference frame that serves to describe the way of partitioning.

Yet, it seems that the question about the uniformity of the convergence of the partitioning is a matter of trivial mathematical argumentation because the postulated in the stochastic thermodynamics large gap between the macroscopic and dynamical time scales suggests that there is "enough time" for performing all dynamical configurations and so the averaged probability converges to the same value at every location. That is why at every location, the integration over all available delays is executed along with the integration over all available frequencies. Though this protocol eventually renders the averaged probability invariant, the current non-uniformity drives a conflict with the stability of the system. Indeed, the non-uniformity permanently propels sharp discontinuities in the spatio-temporal configurations that on exceeding the local thresholds of stability bring about irreversible local defects whose unrestrained development would ultimately result in the system breakdown.

Another argument in favor of the high non-triviality of the subject considered is the relation between the issue of the time scale separation and the velocity anzatz which states that the velocity of transmitting matter/energy/information through every media is always bounded. The account for the boundedness of the velocity of every motion makes the time for performing all dynamical configurations (time of convergence) subject to the size of the system. If so, the response to every perturbation would be intensive variable which renders the stability of the system subject to its size – an obvious conflict with our experience.

To summarize, the issue about the uniform convergence of the averaging protocol brings together the time-translational invariance (principle of relativity) and the stability in a startling puzzle – along with the stability of the system considered, the unrestrained development of local instabilities violates the time-translational invariance of its macroscopic behavior by means of permanent signaling out frequencies in the corresponding power spectrum. And since the presence of each special frequency implies initiation of a specific physical process, the violation of the time-translational invariance implies violation of the principle of relativity.

Unitarity of the Dynamics

A closer examination of the above considered attempts to incorporate the boundedness into the traditional statistical mechanics delineates common misunderstanding that can be subsumed in the following way: all interactions are considered isolated and independent from one another. In turn, as it has been explicitly demonstrated, considerations such as the one mentioned above yield critical violation of the time-translation invariance of the macroscopic behavior of the corresponding system. Indeed, the application of the velocity anzatz, i.e. a restraint over the velocities of atoms, molecules and collective modes, yields unrestrained development of local instabilities which results in an unstable and non-reproducible behavior of the corresponding system. What makes our statement highly non-trivial is that the consideration describing all interactions as isolated and independent from one another commences from the grounding postulate of the traditional statistical mechanics that all interactions are to be subject to unitary evolution. The latter is a key ingredient for providing time-translational invariance of the macroscopic behavior by means of validating the Law of Large Numbers. To remind, the basic assumption of the traditional statistical mechanics consists of associating time-independent frequencies with the development of every macroscopic fluctuation. Therefore the time average on every macroscopic time scale becomes uniquely determined and time-translational invariant. The importance of the latter property is that it provides the reproducibility of the events, both deterministic and stochastic. A decisive pre-requisite for providing time-independence of the frequency associated with development of any fluctuation is the unitarity of the dynamics. The latter implies additive decomposition of each many-body interaction to a succession of independent from one another "clusters" of n − body interactions so that the dynamics of each and every of them to be described by a specific time-independent Hamiltonian whose evolution is set unitary. In turn, this setting justifies the validity of the Law of Large Numbers which states that the sample average of the sum of random independent events asymptotically approaches the expectation value. The major consequence of the validity of the Law of Large Numbers is that it provides the covariance of the transition rates through associating the overall result of the interactions by covariant probabilities (hereafter covariance implies no preferred basis condition, i.e. independence from the choice of auxiliary reference frame); to remind that the expectation value is the averaged over the corresponding phase space collision rate which in turn provides the covariance of transition rates and the state variables.

The latter settings are not fault free. Their flaws can be subsumed by the following considerations:

1. The consideration that all interactions are independent from one another opens the door to arbitrary large local accumulations of matter and energy at any locality and at any time. So far it has been taken for granted that these accumulations are automatically averaged out by means of the Law of Large Numbers alone. However, an overlooked point is that since all local fluctuations are independent from one another, prior to averaging out, any large enough local accumulation of matter and energy can trigger novel, even hazardous for the stability events: phase transitions, pattern formation, sintering, overheating etc. whose major property is that each of them introduces unpredictability and non-reproducibility of the behavior of the corresponding system both on micro and macro-level. In result the system would behave in an uncontrolled, unstable, unpredictable and non-reproducible way. Thus the demand for a stable, controlled and reproducible evolution sets the requisition about existence of a self-sustained *physical* mechanism for permanent control over local accumulation of matter and energy whose action is *prior* to the action of the Law of Large Numbers;

2. The assertion about independence of all interactions from another is only approximation that works well for very low concentrations only: indeed, along with the unitary processes there always are non-unitary processes which come out as a result of the inevitable involvement of the random motion of the species. Any collision could be interrupted by another species that enters the interaction as a result of its random motion. The non-unitarity is due to the time-reversal *asymmetry* between the ingoing and outgoing trajectories of the colliding species which commences from the dependence of the outgoing trajectory on the status of interaction at the moment of arrival of the new species. Thus, the ubiquitous coexistence of unitary and non-unitary interactions reveals an imminent controversy of the traditional statistical mechanics: the

Law of Large Numbers is discarded by the inevitable and ubiquitous presence of non-unitary since it renders collisions not independent from one another.

Therefore our task now is to demonstrate that the concept of boundedness appears as decisive ingredient for providing time-translational invariance of the macroscopic behavior by means of implementing a generic physical mechanism for control over development of local fluctuations. More precisely, we will prove that self-sustained control over local variations of the macroscopic variables is implemented by a generic feedback between the linear-dispersed gapless collective excitations of the macroscopic variables and the non-unitary interactions considered in a setting based on boundedness. We will prove that the major property of that feedback is to provide not only the boundedness of the fluctuations of the macroscopic variables but their covariance as well, i.e. we shall prove that the fluctuations neither select nor signal out any special time and space point. Further, we will prove that the feedback automatically sets the admissible transitions from any given state.

NON-UNITARY INTERACTIONS AND THE INTER-LEVEL FEEDBACK

In the previous section we have demonstrated that the hypothesis of self-organization turns inconsistent under the restraint over the velocities of constituent species and collective modes alone. Thus it seems that either the hypothesis of self-organization or the concept of boundedness is compromised. On the other hand, there is strong evidence that each of these concepts is correct. If so, there must be a way out of this problematic situation and we claim that it lies in a highly non-trivial interplay between the non-unitary interactions and gapless collective modes. Next we will demonstrate that this interplay reconciles the hypothesis of self-organization and the concept of

boundedness by means of substantiating a general inter-level feedback which plays a two-fold role: it restrains the amplitude of local fluctuations at the quantum level and at the same time defines the admissible transitions from any given state at the macroscopic level.

Alongside, thus obtained general physical mechanism for restraining the development of local fluctuation comes to provide novel comprehending of the Law of Large Numbers viewed so far as the only implement for providing time-translational invariance on macroscopic scale.

Mechanism for Controlling Fluctuations

An exclusive generic property of a non-unitary interaction is that its time asymmetry renders each and every such interaction dissipative. Indeed, the time asymmetry renders the asymmetric role of any species that is as follows: the impact of the $i-th$ species on the $j-th$ one does not turn out equal to the impact of $j-th$ species on the $i-th$ one because we have to take into account the non-equal impact on each of them by a third species entering the collision. Thereby the corresponding Hamiltonian turns always into non-Hermitian (its eigenvalues are complex numbers), hence dissipative. On the other hand, we look for a mechanism able to "disperse" every locally accumulated by unitary interactions energy/matter. And namely the generic dissipativity of the non-unitary interactions is a key ingredient of that mechanism. The next step is to establish how the dissipated by a non-unitary interaction energy becomes dispersed throughout the corresponding system. We assume that it happens as follows: the dissipated from non-unitary interaction energy resonantly activates an appropriate local linearly-dispersed gapless mode (e.g. acoustic phonon) of the corresponding state variable (e.g. concentration), by means of which "dispersion" of the accumulated matter over the entire system is achieved. Further, the need for stable and reproducible functioning of a system imposes the following general requirement to the operation of that feedback: it must provide those fluctuations covariant, i.e. it must provide independence of the characteristics of each and every of them from their location and moments of their development.

The concept of boundedness, applied to the operation of the above feedback, implies that it must maintain stability by means of providing the effect of each and every interaction to be always bounded. This implies that whatever the specific characteristics of any given non-unitary interaction are, the operation of the feedback is such that the effect over each and every of the colliding species is finite and bounded. This property delineates the fundamental difference between a unitary and a non-unitary interaction. Indeed, whilst the unitaritiy implies linearity and additivity of the corresponding interactions at the expense of allowing arbitrary accumulation of matter/energy, the boundedness applied to non-unitarity keeps the effect of any interaction bounded at the expense of introducing non-linearity and non-additivity of its execution. Consequently, the non-unitary interactions acquire the following 3 generic properties imposed by the boundedness:

1. The Hamiltonian of a non-unitary interaction is described by a non-symmetric random matrix of bounded elements. The property of being random is to be associated with the interruption of an unitary collision by another species at a random moment; and since the original distance to the interrupting species is of no importance for any given collision, a non-unitary interaction is not specifiable by metric properties such as the distance between species - it is rather specifiable by the correlations among the interacting species so that the intensity of these correlations are permanently kept bounded within metric-free margins.

2. The requisition about limited effect of the interactions over each participating species is further substantiated by setting the wave-functions to be bounded irregular sequences

(BIS). Indeed, an exclusive property of the BISes is that each one is orthogonal to any other; alongside, each of them is orthogonal to its time and spatial derivatives. The latter property is crucial for providing the covariance of the non-unitary interactions since: (i) the orthogonality of every BIS to its spatial and time derivatives provides independence of each and every non-unitary interaction from the interaction path of any participating species; in turn, the obtained independence renders every non-unitary interaction a local event, that is: it depends on current status of participating species only; (ii) alongside, the orthonormality of the BISes renders the probability for finding a species independent from its current position, and from the position of its neighbors. Next we present in details the proof about the above properties of BIS.

3. In order to keep the effect of non-unitary interactions permanently bounded and covariant, we suggest that the transitions introduced by them are to be described by operations of coarse-graining. To remind, the coarse-graining is mathematical operation which acts non-linearly and non-homogeneously on the members of BIS under the mild constraint of maintaining their permanent boundedness alone. As discussed in Chapter 1, this makes a set of BISes dense transitive set where the coarse-graining appears to be an operation that transforms one BIS into another. Thus, the self- consistency of the frame is achieved: the eigenfunctions are supposed always BIS and the transitions described as operations of coarse-graining sustain that property since an operation of coarse graining transforms one BIS into another. But do they compose a complete orthonormal basis in a Hilbert space? The completeness is automatically ensured by the transitivity of the set of BISes. Let us now estimate the orthogonality:

$$I = \int_V \psi_i * \psi_j \, dV \qquad (3.28)$$

Evidently when $i = j$:

$$I = \frac{1}{V} \int_V |\psi_i|^2 \, dV = \sigma^2 \qquad (3.29)$$

where σ is the variance. We already know that every BIS has finite variance. In turn, it ensures the normalisability of BIS as functions that can set an orthonormal basis.

For $i \neq j$ the product $\psi_i * \psi_j$ can be considered as an operation of coarse-graining over either $\psi_i *$ or ψ_j:

$$I = \frac{1}{V} \int_V \hat{G} \psi_j \, dV = \frac{1}{V} \int_V \psi_l \, dV = o\left(\frac{1}{V}\right) \qquad (3.30a)$$

$$I = \frac{1}{V} \int_V \hat{G} \psi_i * dV = \frac{1}{V} \int_V \psi_k * dV = o\left(\frac{1}{V}\right) \qquad (3.30b)$$

where V is the volume of the system and $\hat{G}$ is the operator of coarse-graining that transforms one BIS into another. The estimation $o\left(\frac{1}{V}\right)$ comes from the dynamical boundedness which, to remind, imposes finite distances between the zeroes of every BIS. In turn, the average of every BIS is always of the order of $o\left(\frac{1}{V}\right)$. So, each BIS is orthogonal to every other.

The boundedness and the lack of any steady symmetry of the Hamiltonian that describes non-unitary interactions $\hat{H}_{non}$ renders its eigenfunctions to be always BIS. This assumption is consistent not only with the boundedness but it is consistent also with the general conjecture of the quantum mechanics that after a transition, the

eigenfunction of the final state is superposition of the eigenfunctions at the initial moment. So, we have:

$$\Psi_i = c_{ij} \psi_j \qquad (3.31)$$

where Ψ_i are the final eigenfunction; ψ_j are the initial eigenfunctions. Since $\hat{H}_{non}$ is supposed non-steady and thus permanently induces transitions, its evolution is subject to the time-dependent Schrodinger equation even though it does not involve time explicitly:

$$\frac{\partial \psi_j}{\partial t} = \frac{\hbar^2}{2m} \Delta \psi_j + \left(\hat{H}_{non} - E\right) \psi_j \qquad (3.32)$$

The transition probabilities c_{ij} can be determined through multiplying (3.32) by integrating over the entire volume:

$$c_{ij} \int_V \psi_i * \frac{\partial \psi_j}{\partial t} \, dV =$$
$$c_{ij} \int_V \psi_i * \Delta \psi_j \, dV + c_{ij} \int_V \psi_i * \hat{H}_{non} \psi_j \, dV - c_{ij} E \int_V \psi_i * \psi_j \, dV \qquad (3.33)$$

Next I shall evaluate each term in (3.33) starting with:

$$I_1 = \int_V \psi_i * \frac{\partial \psi_j}{\partial t} \, dV \qquad (3.34)$$

The understanding of the spatial and time derivatives in (3.32) is new: since the operation that preserves boundedness is the coarse-graining, the spatial and time derivatives must be considered as operations of coarse-graining. Let us start with the time derivation $\frac{\partial}{\partial t}$ regarded as operation of coarse-graining. Applied to BIS it commutes with the integration over the volume. In turn this yields:

$$I_1 =$$

$$\int_V \psi_i * \frac{\partial \psi_j}{\partial t} dV =$$

$$\frac{\partial}{\partial t} \int_V \psi_i * \psi_j dV = \qquad (3.35)$$

$$\frac{\partial}{\partial t} a \delta_{ij} = 0$$

where a is the expectation of the BIS that serve as initial eigenfunctions. Since all eigenfunctions are zero-mean BIS, $a = 0$. It should be stressed that $\dfrac{\partial}{\partial t}$ and integration over the volume considered as operations of coarse-graining commute only when applied to BIS!

Similarly:

$$I_2 = \int_V \psi_i * \Delta \psi_j dV = -\int_V \nabla \psi_i * \nabla \psi_j dV = -\delta_{ij} a'$$
$$(3.36)$$

where $a' = 0$ is the expectation value of the gradient of the initial eigenfunctions. The evaluation of the term $\int_V \psi_i * \hat{H}_{non} \psi_j dV$ requires additional supposition about the size of the Hamiltonian $\hat{H}_{non}$. Evidently, the matrix elements $\hat{H}_{ij}$:

$$\hat{H}_{ij} = \int_V \psi_i * \hat{H}_{non} \psi_j dV \qquad (3.37)$$

are non-zero only if the size of $\hat{H}_{non}$ is finite. Then, the equation for the transition probabilities c_{ij} reads:

$$c_{ij} \left(\hat{H}_{ij} - E \delta_{ij} \right) = 0 \qquad (3.38)$$

Equation (3.38) needs additional attention because it has two-fold meaning. On the one hand, it is equation for evaluating the eigenvalues E. On the other hand, it gives the transition probabilities. Though the eigenvalues can be evaluated

for any given $\hat{H}_{non}$, the permanent variations of the matrix elements make this task meaningless. That is why we will focus the attention on the statistical properties of the eigenvalues. The next step is to find their distribution. They are the roots of the characteristic equation of (3.38) that reads:

$$P(E) = \sum_k s_k E^k = 0 \qquad (3.39)$$

where s_k is the coefficients of the polynomial. The most important property of $P(E)$ is that it is random polynomial. This is an immediate result of the variability of the matrix elements G_{ij}. So, it is to be expected that at every instant all coefficients s_k are non-zero. It has been proven (Huges et al, 2004) that the roots of a random polynomial with non-zero coefficients tend to cluster near unit circle and their angles are uniformly distributed.

The clustering of the eigenvalues at the unit circle provides not only boundedness of the amplitude of each fluctuation but boundedness of the rate of development of a fluctuation. Indeed, the rate of a fluctuation is given by the real part of any eigenvalue and thus it is not only bounded but covariant as well; alongside the corresponding rate of its development is given by $\dfrac{extension}{duration}$; the *duration* is given by the imaginary part of the corresponding eigenvalue $\operatorname{Im} E_i$) and its *spatial extension* is also related to the imaginary part of the same eigenvalue trough the dispersion relation of the excited mode; since it is assumed that the feedback operates through excitation of linearly dispersed gapless modes $\lambda = \dfrac{c}{\hbar^{-1} E_i} = \dfrac{c}{\hbar^{-1} \left(\operatorname{Re} E_i + i \operatorname{Im} E_i \right)}$; thus the extension length is $\lambda_{ext} = \operatorname{Im} \lambda = -c \operatorname{Im} E_i$; thereby the rate of development $\left| \dfrac{extension}{duration} \right|$ is proportional to the velocity of the acoustic phonons c. Note that the independence of the rate of development from the size of the fluctuation, from the particularities of the interactions and from the moment and location of its development renders its covariance.

The generalization of the above presented results reads that there exists natural mechanism for maintaining the stability of a system by means of permanent preventing accumulation of arbitrary large amount of matter and energy. The presence of non-unitary interactions along with linearly dispersed gapless modes such as acoustic phonons in every many-body system renders the feedback universally available. We have proved that the operation of that mechanism under boundedness renders fluctuations covariant, extended to bounded size both in space and in time objects which appear, develop and relax with bounded rate. This is an exclusive property of the interplay between boundedness and the non-unitary interactions and thus it is to be expected that its impact on the evolution of the system opens the door to new properties among which is the constitution of a feedback between the quantum and the macroscopic hierarchical levels. Note that the universality of the mechanism for restraining local fluctuations of the matter/energy accumulation to be within given margins gives us credible arguments for considering it as the implement of the inter-level feedback that operates between the quantum and the macroscopic level of matter organization. Indeed, the established above covariance of the bounded fluctuations of the local matter/energy accumulation apparently suggests their association with the instant values of the local macroscopic variables such as concentration and temperature. Thus each of them is bounded to vary so that not to exceed certain specific thresholds.

Self-Organization and the Thermodynamical Limit

A major consequence of the obtained above boundedness of local fluctuations is that it releases the notion of a macroscopic variable from necessity to obey the thermodynamical limit. To remind, the traditional approach to self-organization requires the macroscopic variables which participate to the corresponding reaction-diffusion equations to meet the thermodynamical limit. The latter implies that their values are to be invariant under arbitrary partitioning of the system. It seems that the problem is automatically resolved since the boundedness of local fluctuations definitely provides uniform convergence under arbitrary partitioning. However, the non-triviality of the problem resides in possibility of interference with the persistent presence of specific spatial and temporal scales set by the balance among the processes that proceed in a system. A common example is a reaction-diffusion system where there is at least one pair of spatio-temporal scales set by the balance between the diffusion and the reaction. Indeed, the balance between the surface phenomenon viewed as diffusion and the volume phenomenon viewed as a reaction, selects a specific spatial and time scale where it occurs. Since the selection of those specific scales is a generic property of every reaction-diffusion system, it discards the relevance of the view over the thermodynamical limit as invariance of the macroscopic variables under partitioning. Note, that the ideal gases are the only systems where no specific scales are established. Then, only for them the notion of thermodynamical limit is not controversial.

The successful resolution of the problem exploits the exclusive property of covariance of local fluctuations provided by the mechanism introduced in the previous sub-section. Its key property is that the universality of the feedback between the non-unitary interactions and the gapless modes provides insensitivity of each and every local fluctuation to the specific time- and spatial scale set by the balance between the processes developed in the corresponding system. Put in other words, we exploit the property of the feedback between the non-unitary interactions and the gapless modes to operate in the same way irrespectively to the details and nature of the corresponding unitary interaction. In this line of reasoning it is worth noting that whilst the unitary interactions are specific for every system, the feedback between the non-unitary interactions and the gapless modes operates in a uniform way. Thus, the covariance along with the boundedness of the local fluctuations justifies the introduction of macroscopic variables avoiding the need for holding the thermodynamical limit.

Further, it is worth noting that our approach assigns two specific pairs of time- space scales to every system: the first one is to be associated with the balance between the corresponding processes; the second pair is to be associated with the thresholds over the size and duration of the fluctuations. To remind, the size and duration of every fluctuation are related through their association with the real and the imaginary part of the corresponding eigenvalue (see the considerations after (3.39)). Then, it is to be expected that the system has its own internal metrics which is created and substantiated by the dynamics. Since the metrics arises spontaneously, the avoidance of the thermodynamical limit acquires novel aspect of understanding: its holding is not regarded any longer as a necessary condition for thermodynamical description of self-organized systems in the traditional setting. In fact according to the traditional statistical mechanics its holding is necessary for equivalence between micro-canonical and canonical description of a system. However, neither of these approaches pre-supposes metrics; on the contrary, in both approaches it is supposed that the micro-states are described by probabilities while transitions between any two of them are arbitrary. Thus, it is impossible to define any metrics in a state space where arbitrary transitions are allowed. Therefore, the avoidance of the thermodynamical limit implies that we are free to build a novel approach where metrics appears naturally as a defined property.

The existence of a feedback between the non-unitary interactions and the gapless modes is a decisive implement for resolving the long-standing paradox between the dynamical reversibility and the macroscopic irreversibility. The paradox consists of the following controversy: on the one hand, the assumption that all interactions are unitary, hence subject to time-reversal symmetry, renders a non-preferred direction for their evolution. On the other hand, our daily experience explicitly shows that macroscopic systems spontaneously choose only one direction for their evolution: ink dissolved in the water is uniformly distributed in it and never deviates from the achieved uniformity. This is the simplest example for the necessity to renounce the idea of the dynamical time-reversal symmetry as property of all interactions. This constitutes our present task: to find out how the feedback between the non-unitary interactions and the gapless modes helps to resolve the paradox. So far we have associated the need of such feedback with the demand for sustaining a bounded variability of the macroscopic variables only. We have proved that the established in the sub-section *Mechanism for controlling fluctuations* physical mechanism for maintaining their boundedness is explicitly grounded on the generic property of the non-unitary interactions to be time-irreversible. Thus time-reversal symmetry stands as a property of the unitary interactions only whilst its breaking for the non-unitary interactions renders the latter dissipative; therefore their role is to participate into a feedback along with gapless modes so that to "disperse" throughout the system the extra matter/energy which is locally accumulated by the unitary interactions. Further, since the 'dispersion" happens wherever and whenever an extra matter/energy is locally accumulated, the ink is always uniformly distributed in the water; then any visible deviations from the uniformity are *impossible*. It is worth noting that the ban over reversibility constitutes the major difference with the traditional statistical mechanics where, though macroscopic reversibility is considered highly improbable, it is still *possible*.

CONCLUSION

In the present chapter an introduction was made to the concept of boundedness on quantum level in a self-consistent way. The difficulties in definition of this concept at quantum level come out from the fact that, to the most surprise, the velocity anzatz alone not only does not provide automatic damping of the local fluctuations but, on the contrary, it brings about their amplification. In result, not only the system's stability is seriously threatened, but it is impossible to define the notion of a macroscopic variable. Indeed, a macroscopic variable, such as a concentration and temperature requires independence from the size and shape of the volume where it has been defined. The latter is achievable if and only if the local instabilities are bounded. However, what appears to be the most natural way of introduction of boundedness, namely through putting restraint over the variability of the velocities, yields surprising amplification of the local instabilities. Further, we prove another seemingly natural way of introducing boundedness, namely through dissipation of the locally accumulated matter/energy by means of resonant activation of specific for the system collective modes. And again it turns unlikely that the dissipation of the locally accumulated matter/energy happens in a uniform way.

Thus we face the question whether it is ever possible to expect that the boundedness is a credible concept for the quantum phenomena. The negative answer to this question may come by accepting the view about non-relevance of the boundedness at quantum level. If so, however, it would open the door to a cascade of problems the core of which would be the withdrawal of the boundedness as a principle of organization at every level of matter/energy organization. Further, we would loose the general implement for substantiating the idea about self-organized semantics as a generic property of the intelligent complex systems. Moreover, the obtained by us results put under threat even the traditional notion of a macroscopic variable viewed as a variable that is insensitive to the size and shape of the volume where it is defined. In turn, it would be impossible to expect any multi-level hierarchical organization since it is impossible to define the macroscopic variables as independent from the dynamical details of the constituent atoms and molecules. Last but not least, it would be impossible to define any form of a law, viewed as a relation between certain variables so that this relation to be independent from the choice of the reference frame.

There is a way out of this situation and boundedness is available at the quantum level. Its introduction turns out self-consistent when considering the role of the non-unitary interactions which along with the unitary ones are always present as a part of all interactions. Taking into account that the non-unitary interactions are dissipative because they do not obey time-reversal symmetry, we assume that the locally accumulated extra matter/energy dissipates through the following mechanism: the non-unitary interactions in any local fluctuation dissipate through resonant activation of the local gapless modes (e.g. acoustic phonons). The central idea related with the introduction of the boundedness is the assumed by us replacement of operation linear superposition with operation of coarse-graining. To remind, the operation of coarse-graining is a non-linear and non-homogenous operation which preserves boundedness while the generic for the traditional quantum mechanics operation of linear superposition allows local accumulation of arbitrary large amount of matter/energy. We have proved that the proposed by us mechanism indeed achieves automatic control over the local fluctuations: indeed, any extra accumulation of matter/energy produced by the unitary interactions is automatically damped through the coupling between the non-unitary interactions and the acoustic phonons. It is worth noting that the intensity of the non-unitary interactions increases on an augmentation of the locally accumulated matter/energy which in turn intensifies the dissipation. Thus, the larger the fluctuation the faster it dissipates.

A very important property of the coupling between the non-unitary interactions and the acoustic phonons is the covariance of the corresponding dissipation viewed as insensitivity to a particular choice of the reference frame. This is exclusive property of the boundedness, implemented by the coarse-graining, which renders the current state of any atom (molecule) independent from the interaction path. This property verifies our assumption that the non-unitary interactions are metric free in the sense that their participation in the corresponding Hamiltonian is not specified by metric properties such as distances, velocities etc. Yet, the apparent dependence on the number of surrounding species prompts the suggestion that it is controlled by the current local concentration. Thus, the inter-level feedback becomes evident: on the quantum level it controls the local fluctuations so that each of them stays bounded at any moment; thus the concentration is kept permanently well defined; in turn, the well defined concentration sustains and controls the intensity of the non-unitary interactions and thus put the local accumulation of matter/energy subject to boundedness.

What is important for the self-consistency of our theory is to establish whether the proposed mechanism for the control over the local fluctuations is the desired inter-level feedback. In other words, we must consider whether it successfully develops the idea of self-organization so that on the one hand to eliminate the problems with the local instabilities and on the other hand to appear as an inter-level feedback whose generic properties are the presence of a continuous band of the shape $1\big/f^{\alpha(f)}$ in the power spectrum and, along with a discrete band, the presence of an non-recursive frequency line in the same power spectrum. In the present Chapter we considered the problem about the control over local instabilities. In the next Chapter we will consider the role of the controlling mechanism in substantiating the inter-level feedback with the desired properties.

REFERENCES

Gardiner, C. W. (1985). *Handbook of stochastic methods for physics, chemistry and the natural science* (Haken, H., Ed.). *Vol. 13*). Berlin, Germany: Springer Series in Synergetics. doi:10.2307/2531274

Hughes, C. P., & Nikeghbali, A. (2008). The zeroes of random polynomials cluster uniformly near the unit circle. *Compositio Mathematica, 144*, 734-746. Retrieved from http://arXiv.org/abs/math.CV/0406376

Chapter 4
Hierarchical Order II:
Self–Organization under Boundedness

ABSTRACT

The self-organization under boundedness is considered as an operational protocol, which describes the highly non-trivial interplay between the inter-level feedback and the spatio-temporal pattern that describes the corresponding homeostasis. A generic property of the feedback is that it sets metrics in the state space and defines the admissible transitions from any given state. Further, the state space is partitioned into basins-of-attraction so that each of them is tangent to the point called accumulation point. A distinctive property of the basins-of-attraction is that the discrete band of the power spectrum appears as intra-basin invariant and thus turns as appropriate candidate for a "letter." The accumulation point is associated with the notion of a "space bar." Then, since the motion in a bounded attractor is orbital, it is appropriate to associate a "word" with an orbit. The latter open the door for assigning a specific non-mechanical engine to each and every orbit. In turn the functional irreversibility of any engine substantiates the sensitivity to permutations of every semantic unit.

INTRODUCTION

The theory developed in the present book considers hierarchical order as the major implement for strengthening the response through its specification that happens at different hierarchical levels.

Thus, on the one hand, the hierarchical structuring serves as an implement for enhancing stability of a complex system. On the other hand, the specification of the response viewed as response to different external stimuli at different hierarchical levels poses the question whether the variety

DOI: 10.4018/978-1-4666-2202-9.ch004

of the responses must obey a general rule for its organization. Moreover the following question arises: how to specify the differences in the response at different hierarchical levels.

It is obvious that a multi-level hierarchical organization implies generic insensitivity of the level response to the variables by which the response is described at any other level. This consideration strongly suggests formation of collective variables for every level of hierarchical organization. On the other hand, the very idea of hierarchical organization implies that each level "feels" the others, i.e. the levels are interconnected by feedbacks. If otherwise, the system would not be a single complex structure but would fall apart into several simple disconnected systems. These considerations suggest that the presence of inter-level feedbacks is a necessary part of the hierarchical organization of any complex system. Thus the question that arises is how to take into account the inter-level feedbacks so that even though each of them commences at a different hierarchical level, it is described by the variables specific for a given level.

The boundedness gives a new prospective to the inter-level feedbacks: they serve to control the amplification of the local fluctuations of the collective variables that describe a given level. As we have considered in the previous Chapter, the amplification of the local fluctuations is a generic property of all systems made of atoms and molecules, i.e. a generic property of systems whose constituents are allowed to exert free motion, even when its velocity is bounded. This fact initiates a systematic study of the roots of the problem and they were found at the quantum level. The root of the problem turns out to be the so far over-looked interplay between the random and the potential motion of the constituents which results in non-unitary interactions. Then, the consideration of the non-unitary interactions under the concept of boundedness provides their exclusive property to be metric-free. The latter implies that their characteristics are not specified by metric properties such as position, velocity etc. The property of being metric-free is justified by the properties set by obeying the operation of coarse-graining instead of that of linear superposition. To remind, the operation of coarse-graining, introduced in Chapter 1, is a non-linear operation which acts non-linearly and non-homogeneously on a bounded irregular sequence so that to preserve its boundedness; on the other hand, the linear superposition allows accumulation of arbitrary amount of matter/energy locally and thus is inconsistent with the concept of boundedness. Although the non-unitary interactions are metric-free, their intensity is controlled by the local concentration of the species.

Thus, we become able to outline the feedback: the non-unitary interactions control local fluctuations of the concentration by permanently keeping the size and the rate of development of each fluctuation bounded. Alongside, the local concentration permanently controls the intensity of the non-unitary interactions. Thus, we come to the conclusion that the local concentration permanently varies within specific margins. Yet, the following questions remain to be clarified: what is the role of the unitary interactions and how does the interplay between them contributes to the hierarchical organization?

The goal of the present Chapter is to demonstrate that the interplay between the unitary and non-unitary interactions gives a new life to the idea of self-organization making it central for the hierarchical self-organization. The novelty consists of giving credible grounds for substantiating the idea of incorporating the inter-level feedbacks as a generic property of the self-organization. Moreover, we will demonstrate that the thus defined self-organization opens the door to semantic-like organization of the hierarchical order by means of substantiating the physical implementation of an "alphabet" and a "space bar." It is worth noting that the natural implementation of a "space bar" is an essential pre-requisite for any form of semantics: it serves as a natural separator of the words (semantic units) from one another. To compare, the traditional computing uses hand-craft resetting to "neutral" position after each step in the algorithm. Thus, the existence of a "space-bar" renders corresponding semantics autonomous, i.e. its comprehension does not require any "external mind" for its reading. Moreover, we shall demonstrate that the re-defined theory of self-organization opens the door to a two-fold presentation of a semantic unit: through a sequence of letters and through the corresponding orbit in the state space. The crucial advantages of this presentation will be among the central issues of the rest of the book.

BACKGROUND

The self-organization is one of the most advanced and fascinating hypothesis about the behavior of extended many-body systems with short-range interactions. It asserts that emergent properties on macroscopic scale, obtained as a result of self-organization, are insensitive to the microscopic dynamics and are independent from the individual elements and components. The mild conditions for derivation of the reaction-diffusion equations (hyperbolic partial differential equations) that serve as mathematical tool for its description support the ubiquity of its application: the idea of self-organization is applied to wide spectrum of systems as diverse as chemical systems, optical lattices, social systems, population dynamics etc. Indeed, these equations are founded on the postulate that the diffusion does not create or annihilate substance which provides the conservation of substance flow in space-time.

However, the rigorous derivation of the reaction-diffusion equations requires explicit demonstration that the conditions at which the microscopic dynamics does not contribute to the emergent properties are as credible as the idea of flux conservation. The considerations in the previous Chapter explicitly demonstrated that the microscopic dynamics actually contributes to the emergent properties by means of involving the

bounded fluctuations of the collective variables. Thus, we face the dilemma whether this implies a change of the entire concept of self-organization or it is just a technical fact of incorporating the inevitable fluctuations in every specific case. The importance of the dilemma consists of the fact that while the first alternative renders self-organization as the major functional tool for the hierarchical self-organization in a semantic-like way, the second alternative allows considering the self-organization only as a phenomenon, not as an operational protocol. To make this point clearer, it is worth noting that the phenomenon is considered as setting a specific relation between variables characterizing it at specific pre-determined environment; at the same time an operation protocol implies that self-organization exhibits certain generic properties which are invariant in a non-specified environment. In the present Chapter we will give credible basis to the first alternative, i.e. we will prove that the self-organization is the operational protocol for semantic-like multi-level hierarchical self-organization.

The crucial argument in favor of the first alternative is the proof that the generic property of the solution of the corresponding reaction-diffusion equations is the additive decomposability of its power spectrum to a continuous band of shape $1/f^{\alpha(f)}$ and a discrete band which comprises a non-recursive component along with the principal one. Thus, we proved that the present modification of the idea of self-organization not only preserves its original goal, namely to give rise to emergent phenomena, but it acquires a new quality, namely it appears as generic operational protocol for substantiating hierarchical order. Besides, the presence of non-recursive component in the power spectrum implements a non-extensive diversification of the hierarchical self-organization. By this we mean that different states (levels) characterized by different non-recursive lines are algorithmically unreachable from one another. These considerations render enormous importance of the hierarchical self-organization of intelligent systems making the questions about defining any extensive criterion for intelligence meaningless. To make it clearer, we assert and we will prove that the idea of semantic-like non-extensive multi-level hierarchical self-organization automatically discards any extensive criterion for information content and/or intelligence; thus neither the quantity of bits, nor the size of a brain, nor the number of letters in an alphabet is decisive for the capacity of an intellect. Thus, we assert that intelligence is not quantitatively measurable; yet, following our experience, we assert that the more organized hierarchy the more complex its intelligence is.

EQUATION-OF-STATE

The viewpoint that the hierarchical order implements strengthening of the response by means of its specification and diversification demands that any state at any level of the multi-level hierarchy to be defined by specific to it control parameters. Put it in other words, this implies that the state space of the system is metric and is enumerated in a specific way. This is radically different approach compared to the traditional statistical mechanics where the state space is probabilistic and where no metrics is assigned. Indeed, since the transition among states is supposed to be a Markovian one, i.e. each state can follow and/or precede any other, the only possible description is the probabilistic: we can define only the probability by which a system occurs in any given state.

However, the concept of boundedness definitely does not fit the statistical approach since it automatically discards the Markovianity of the transitions. Indeed, let us suppose that the transitions obey the Chapman-Kolmogorov relation which implies that the probability for the next transition depends only on the current state. The question is whether this necessary for establishing Markovianity condition is also the sufficient one. Next we demonstrate that it is not.

Let us suppose that the transition from state j to neighbor state i depends only on whether the transition to j has happened. However, the transition to j depends on whether it comes from the range of the admissible states, i.e. those ones that does not violate the boundedness – let us denote them by k. Hence the transition to i is set on the chain of the previous transitions $...l...kj$. Therefore, can the process be non-Markovian when the Chapman-Kolmogorov relation holds?! It seems Markovian because the transition from j to i depends only on j. However, it is non-Markovian, because any admissible transition depends on the succession of the previous ones. (Examples of non-Markovian chains that fulfill Chapman-Kolmogorov relation are presented in (Feller, 1975). Thus, it turns out that the Chapman-Kolmogorov relation is insufficient to provide Markovianity. The present considerations demonstrate also that the boundedness introduces non-Markovianity as a generic property.

Another controversy on our way to re-define the idea of self-organization comes out from the following consideration: in the previous Chapter we established the boundedness of the local fluctuations. Then, it seems that we can conclude that they are averaged out and the result is that their contribution averaged over larger areas is negligible. Thus, we would prove the idea of self-organization as a phenomenon and discard the view that the local fluctuations actually substantiate the idea of the inter-level feedback. The disproval of this consideration is grounded on the highly non-trivial role of the following two facts: the first one is that the averaging out is not legitimate because it is permanently interfered by the presence of at least two spatio-temporal scales; one imposed by the balance between the reaction and the diffusion and the other by the size and rate of development of the local fluctuations. The second argument against averaging out comes out from the metric-free property of the non-unitary interactions: since neither of them depends on distances and velocities but they are controlled by the correlations among distant species (local concentration), it is obvious that the fluctuations are defined as bounded at the same local area where the concentration itself is defined. This substantiates the idea of the covariance of the local fluctuations, i.e. their consistence with the principle of relativity (neither of the fluctuations initiates a specific physical process). To compare, the interplay between averaging out and the presence of at least two spatio-temporal scales actually implies violence of the principle of relativity since the averaging out would eliminate the spatio-temporal scale associated with the fluctuations. Then the question is: what is the physical process that drives its elimination at the larger scales and leaves it at the finest scales?!

Now we are ready to define the generic mathematical object which describes self-organization along with an inter-level feedback. It must be a system of autonomous ordinary or partial differential equations for the mass/energy balance of processes involved; there also must be present the fluctuations induced by the inter-level-feedback. The mass/energy balance is set by the contribution of the corresponding unitary interactions and follows the traditional approaches. The novelty comes from the fluctuations. Let us present as an example the simplest form of the reaction-diffusion equations with an inter-level feedback involved:

$$\frac{d\vec{X}}{dt} = \vec{\alpha} \bullet \hat{A}\left(\vec{X}\right) - \vec{\beta} \bullet \hat{R}\left(\vec{X}\right) + \vec{\mu}_t\left(\vec{X}\right) \quad (4.1)$$

where $\vec{X}$ is the vector of the concentrations; $\vec{\alpha}$ and $\vec{\beta}$ are the control parameters; $\hat{A}\left(\vec{X}\right)$ and $\hat{R}\left(\vec{X}\right)$ are the mass/energy exchanged in the processes taken at that hierarchical level and $\vec{\mu}_t\left(\vec{X}\right)$ are the fluctuations of the corresponding

concentrations induced by the non-unitary interactions (inter-level feedback).

In general Equation (4.1) are non-linear differential equations. It is well established that their solutions do not follow the control parameters in a continuous and uniform way as it is with their linear counterparts; instead the generic property of the non-linear autonomous differential equations is that the control parameter space is partitioned into domains, called basins-of-attraction, so that a specific solution is associated with each basin-of-attraction. Among the admissible solutions are oscillatory ones even when the system is autonomous, i.e. it is not subject to external periodic driving. These oscillations are called self-sustained. Moreover, the power spectrum of the self-sustained oscillations appears as intra-basin invariant. Thus, now we have clear idea how the discrete band in the power spectrum appears and its appropriateness for characterizing the homeostasis. Now we must answer the question about the contribution of the continuous one.

In order to prove that the power spectrum of the solution of (4.1) is additively decomposable to a discrete and the continuous band let us consider the equation for the averaged over a window of length T solution.

$$\frac{d\bar{\vec{X}}}{dt} = \frac{d}{dt}\lim_{T\to\infty}\frac{1}{T}\int_0^T \vec{X}\left(t\right)dt = \lim_{T\to\infty}\frac{1}{T}\int_0^T \frac{d\vec{X}\left(t\right)}{dt} =$$

$$= \lim_{T\to\infty}\frac{1}{T}\int_0^T \left(\vec{\alpha}\bullet\hat{A}\left(\vec{X}\right) - \vec{\beta}\bullet\hat{R}\left(\vec{X}\right)\right)dt +$$

$$\lim_{T\to\infty}\frac{1}{T}\int_0^T \left(\mu_t\left(\vec{X}\right)\right)dt$$

$$(4.2)$$

The term

$$\lim_{T\to\infty}\frac{1}{T}\int_0^T \left(\left(\mu_t\left(\vec{X}\right)\right)\right)dt \to 0 \quad (4.3)$$

since the fluctuations form a zero-mean BIS.

The system of equations:

$$\frac{d\vec{X}}{dt} = \vec{\alpha} \bullet \hat{A}\left(\vec{X}\right) - \vec{\beta} \bullet \hat{R}\left(\vec{X}\right) \qquad (4.4)$$

determines the "homeostatic" response and gives rise to a discrete band in the power spectrum whenever the corresponding solution is self-sustained oscillations. The calculation encapsulated in (4.3) also displays that the "homeostatic" response given by (4.4) is exclusive property of boundedness. Indeed, it comes by the fact that (4.3) holds only for zero-mean BIS. Thus bounededness turns out crucial for the additive decomposability of the response to a "homeostatic" and noisy part viewed as a generic property.

Now it is clear that the power spectrum of the solution of (4.1) is additively decomposable to a discrete band, which comes from (4.4) at appropriate control parameters, and a continuous band of shape $1/f^{\alpha(f)}$, which comes from the inter-level feedback. The fact that both bands in the power spectrum comes from the same equations (4.1) is justified by the presence of a non-recursive line coming from their interplay. The interplay commences from the non-linear and non-homogeneous way of participation of the "unitary" and "non-unitary" components in (4.1). And since the overall solution of (4.1) is bounded irregular function it could be presented as wild and permanent bounded "twisting" and "winding up" around a "skeleton" built on the discrete band of the power spectrum. Thus, the motion resembles the motion on a torus with irrational frequencies. In the present case the motion never stops and thus extra-line associated with irregular motion around the skeleton appears always as non-recursive even-though the thresholds over the corresponding fluctuations are free to be in any relation with the other scales in system.

Thus, the conclusion made is that the response is a two-component variable: the first component is specific to the level while the second one retains universal characteristics. Further, the specific part appears as an intra-basin invariant and thus it is robust to all intra-basin fluctuations. That is why we associate the specific part with homeostasis while the universal one is associated with ever-changing environment. Yet, the variety of responses viewed as self-sustained oscillating output makes possible diversification of the inputs as well by means of using a given output as input for another item in a circuit. It is worth noting that self-sustained oscillations viewed as input gives rise to a new protocol: driven oscillations. In turn, the interplay between self-sustained and driven oscillations along with the impact of the inter-level feedbacks opens the door to a large variety of phenomena. Yet, no matter how complex is the response, its power spectrum is always decomposable to a specific discrete band and universal noise one so that the discrete band always comprises at least one non-recursive component. It should be stressed that self-sustained oscillations are exclusive generic property of the non-linear autonomous differential equations and thus the spontaneous "emergence" of an oscillating pattern from a homogeneous solution justifies the notion of self-organization. Alongside, it should be stressed on another aspect of the self-sustained oscillations: a non-recursive line appears only in a power spectrum with non-zero discrete band. Indeed, the power spectrum of zero-mean BIS comprises only continuous band of shape $1/f^{\alpha(f)}$. Thus, only different "emergent" patterns are algorithmically unreachable from one another. The homogeneous systems appear as algorithmically equivalent thereby requires different prospective for their study.

Last but not least both the diversification of the response and the input achieved through diversification of the discrete band obey boundedness and reproduce it: thus bounded response gives rise to bounded input which in turn produces bounded response etc.

METRICS IN THE STATE SPACE

One of the decisive arguments in favor of the view that the self-organization stands as operational protocol rather than as a phenomenon will come from the consideration that the stochastic terms in Equation (4.1) participate not only as inevitable fluctuations but that they set the admissible transitions from the given state. This task is highly non-trivial and goes through a number of steps. The first one is the presentation of Equation (4.1) in the following form:

$$\frac{d\vec{X}}{dt} = \vec{\alpha}_t^{eff} \bullet \hat{A}\left(\vec{X}\right) - \vec{\beta}_t^{eff} \bullet \hat{R}\left(\vec{X}\right) \qquad (4.5)$$

where $\vec{\alpha}_t^{eff}$ and $\vec{\beta}_t^{eff}$ are effective control parameters chosen so that the instant solution of Equation (4.5) to match exactly the corresponding instant solution of Equation (4.1). Thus, at every moment the solution of Equation (4.5) deviates from the solution of Equation (4.4). The generic property of these deviations is their boundedness: to each point in the control parameter space is assigned certain threshold over the deviations which cannot be exceeded. Thereby the deviations appear rather as a random choice of a single selection of a bounded multi-valued function. The selections of that multi-valued function determine the set of admissible transitions from any given state. In order to prove that the self-organization appears as an operational protocol for inter-level control rather than a phenomenon we must consider whether the properties of the state space allow such consideration. The issue of primary importance is the issue whether the state space retains a metrics. This question naturally arises from the idea that the setting the admissible transitions automatically implies defining distance to any of them. Next we proceed with presenting credible arguments in favor of assigning metrics to the state space.

The correspondence between Equation (4.1) and (4.5) suggests one-to-one correspondence between the control parameter space and the state space. A strong argument in favor of this suggestion is brought about by the established above correspondence in matching the admissible transitions: the shift of the control parameters (Equation (4.5)) is matched through the deviation of the state itself (Equation (4.1)). Keeping in mind that the control parameter space is always enumerable and thereby it always retains certain metrics, the established one-to-one correspondence between the control parameter space and the state space is a crucial argument for assigning metrics to the state space.

Further, we will prove that the generic property of this metrics is that it is locally Euclidian. We start with reminding that the Euclidian metrics is strongly grounded on the notion of the nearest neighbor. A necessary condition for existing of metrics is the possibility for single-scale Voronoi tessellation of the corresponding space; and when each and every Voronoi cell comprises the current nearest neighbors, the metrics is locally Euclidian. To remind that Voronoi tessellation implies a single-scale partitioning of the space into cells so that the latter to be densely packed. Taking into account that every path in the state space under boundedness is realized on a latticised subset (whose details vary from one sample lattice to another), each latticised subset is equivalent to a sample of Voronoi tessellation. Keeping in mind that each cell in every sample of this Voronoi tessellation comprises the current nearest neighbors, we conclude that indeed a state space under boundedness retains metrics and that this metrics is locally Euclidian. It is worth noting once again that the metric is locally Euclidian because the property that each Voronoi cell comprises the current nearest neighbors is exclusive property of the boundedness. Put it briefly, our proof asserts that a set of lattices each of which is equivalent to the same scale Voronoi tessellation, constitutes a continuous space with locally Euclidian metrics.

And any particular path selects its unique latticised subset on which it moves.

Another important aspect of the developed so far approach to the assigning metrics to the state space concerns the requirement about the unique way of defining the nearest neighbors. This question arises from the ambiguity brought about the choice of units by means of which the reaction-diffusion equations are cast in dimensionless form. The difficulty comes from the following well established fact: the partitioning of the control parameter space into basins-of-attraction for the non-linear differential equations depends explicitly on the size of the terms involved in the equations; and the size of the terms explicitly depends on the choice of units. Next we will elucidate this problem in details.

The fact that there are properties which explicitly depend on the choice of units strongly opposes the widely accepted so far viewpoint that the processes in the Nature must be independent of the choice of units. The traditional viewpoint has been taken for granted and has been considered as a common sense. Little attention has been paid to the problem and even it has not been recognized as a problem at all. However, it is a serious problem. Moreover, it turns out to be a fundamental problem of modern mathematics that involves nonlinearities. To resolve the puzzle let us first elucidate explicitly why this problem has not been recognized for more than a century. Let us consider a system of linear ordinary differential equations:

$$\frac{d\vec{x}}{dt} = \hat{A} \bullet \vec{x} \tag{4.6}$$

where $\hat{A}$ is the matrix of the coefficients that determines the eigenvalues of the solution. Obviously, the rescaling of the variables $x_1, x_2, ..., x_N$ by any vector of parameters $(c_1, c_2, ..., c_N)$ leaves (4.6) invariant. Note, however, that the invariance of (4.6) with respect to the choice of units is property of the linear differential equations only! Yet, because of the dominant role of the linear mathematics in all fields of science for more than a century, the problem with the units has not been recognized at all. Only very recently the intensive study of the nonlinearities has attracted the research focus and their adequate incorporation in our knowledge about Nature. It turns out that unlike their linear counterparts, the non-linear differential equations are *not* invariant under the rescaling of the units. Furthermore, an already established general result is that the type and the properties of the solution of the non-linear ordinary or partial differential equations strongly depend on the concrete values of their parameters. To illustrate the point let us consider the following example:

$$\frac{dx}{dt} = ax - b\left(x - k\right)^3 \tag{4.7}$$

where a, b and k are parameters. Obviously, the solution of (4.7) strongly depends on whether $\left(\left(ax - b\left(x - k\right)^3\right)\right)$ is positive or negative. If positive, the solution is unstable and it exponentially departs from the steady one determined by $ax - b\left(x - k\right)^3 = 0$. On the contrary, in the range of x where $ax - b\left(x - k\right)^3 < 0$, the solution is stable and monotonically approaches the steady one. Hence, the demarcation between the stable and unstable solution is determined by:

$$ax - b\left(x - k\right)^3 = 0. \tag{4.8}$$

Let us now rescale the variable x according to $x' = 2x$. Then (4.7) becomes:

$$\frac{d2x}{dt} = 2ax - b\left(2x - k\right)^3 \tag{4.9}$$

Simple calculations show that the equation for demarcation between the stable and the unstable solution becomes:

$$ax - \frac{b}{2}\left(2x - k\right)^3 = 0 \, . \qquad (4.10)$$

Obviously (4.10) deviates from (4.8) in its non-linear part. Therefore, the non-linearities render the differential equations non-invariant under the choice of units. Thus, the choice of units turns out crucial for the non-ambiguous defining the basin-of-attractions in the control parameter space.

Further, the choice of units puts under threat the entire proof of assigning metrics since if there is no way of non-ambiguous way of setting units, the nearest neighbors would vary from one choice to another thereby compromising the idea of a single-scale Voronoi tessellation.

The approach that comes at rescue is the utilization of the natural spatio-temporal scales set by the balance between reaction, diffusion and the thresholds over the size and the rates of fluctuations. We have 2 natural space and time standards: the first one is the scale set by the balance between the reaction and the diffusion; the second one is set by the threshold over the size of the fluctuations. Yet, alone these scales define only the ratio between them. The decisive implement for setting units for space and time separately comes from the threshold over the rate of developing the fluctuations. Its well defined value (the velocity of the acoustic phonons) makes the choice of units non-ambiguous. Thus, for every process there is natural and un-ambiguous choice of units. In turn, this justifies in a self-consistent away our idea of assigning metrics to the state space.

It should be stressed that the assigned metrics to the state space is a dynamical property, i.e. it appears only on exerting bounded motion in the state space. Thus, figuratively saying, the metrics is a property of the bounded motion itself and does not exist independently from the functioning of a system. This makes it radically different from the notion of metrics in mechanics where it is a pre-requisite of the pre-determined space-time viewed as "container" for the matter. The association of metrics with the dynamics is closer to the idea of curved space-time in the general relativity where the local curvature is defined by the local mass. Yet, there is an essential difference: while the local curvature is locally unbounded, the metrics of the state space is always locally Euclidian. In result, no analogs to a "black hole" are available in a bounded state space.

Summarizing, the conclusion made is that the state space of self-organization viewed as an operational protocol of hierarchical structuring retains locally Euclidian metrics and thus provides self-consistent one-to-one correspondence with the control parameter space. Now we are able to answer the question where the bounded environment comes from. Indeed, looking at Equation (4.5) it becomes obvious that they could be considered as a description of a phenomenon put in a bounded environment. What constitutes the difference between this view and self-organization viewed as operational protocol is that association of the fluctuations with inter-level feedback makes the choice of units un-ambiguous. Indeed, the presence of at least two natural spatio-temporal scales along with the velocity of the acoustic phonons appears as the decisive implement for the purpose. On the other hand, when considering the self-organization as a phenomenon in bounded environment the only generic scales are those set by the reaction and diffusion. Therefore, in this case we cannot assign metrics to the state space in an un-ambiguous way. The major disadvantage of the state space without metrics is that it does not allow further hierarchical super-structuring. The most evident argument for this assertion is that any hierarchical super-structuring requires enumeration of its building blocks so that to be able to put them in a certain order to obtain an "orchestra." Thereby the metrics is a decisive property of a hierarchically organized state space.

STRUCTURE AND TOPOLOGY OF THE STATE SPACE

So far we have established that self-organization of open reaction-diffusion systems meets the major property of the complex system: its response is persistently separated into specific and universal part so that the specific part is robust to intra-basin fluctuations while the universal part appears as BIS and thus is characterized by 3 invariants whose major property is insensitivity to the statistics of the fluctuations. We prove more than that: we have modified the idea of self-organization so that it to become the major generic implement for hierarchical structuring. We have proved that the self-organization involves inter-level feedbacks in the form of bounded non-specified environment. The thin line that separates between an "environment" external for the system and an "inter-level feedback" is that the latter provides non-ambiguous setting of the units and provides that all units are expressed through intrinsic for the system spatio-temporal scales. To compare, since setting of the units inevitably involves thresholds of stability and rate of development of the fluctuations, it is obvious that when they come from an "environment" they would change the behavior of the system by means of changing its basins-of-attraction. Then, such system would be too "flexible" for retaining identity. However, such systems are of no interest for us. Thereby we will focus our further attention to systems which are hierarchically self-organized.

By partitioning the state space into basins-of-attraction, a hierarchically organized system acquires a new property: adaptation. Indeed, according to the control parameters, its "homeostasis" resides in the corresponding basin-of-attraction. Thus, loosely speaking, the system "adapts" to current conditions. It is important to stress that the adaptation comes in two ways: the first one is directly via changing control parameters; the second one goes via modification of the inter-level feedback properties. It is worth noting that the two ways of system's submission to adaptation are not independent from one another; on the contrary, their action must be self-consistent. Thus we proved our assertion that hierarchical structuring is an implement for strengthening the response and the adaptation is a new tool for this purpose. To compare, the traditional statistical mechanics explores the idea of a single equilibrium viewed as a global attractor for all initial conditions. This implies that it considers only one state at which a system can reside arbitrary long time for every environment. Thereby, it is totally incompatible with the idea of adaptation. The radical novelty of the idea about adaptation of homeostasis is that it replaces the idea of thermodynamical equilibrium at the earliest stage of hierarchical organization. So far we have considered the simplest two-level system (reaction-diffusion system described by Equation (4.1)) and it turns out that the adaptation appears naturally.

Yet, the simplest systems are not able to reveal the full potential of hierarchical organization. These systems are open spatially homogeneous systems; their state space comprises only two basins-of-attraction: stable steady state and self-sustained oscillations. It is naturally to suggest association of the corresponding intra-basin invariants with information symbols. The obtained result is a 2-bit processor which works as a Turing machine. Does it mean that our dream for a self-organized semantics failed? No, it is not.

The clue is that self-organized semantics is possible only when the state space has certain minimal structure and topology. And the central point is that it must comprise at least 3 basins-of-attraction and an accumulation point. The reason for this is as follows: fundamental ingredient of any semantics is the space bar, which serves as a separator among words. What could be the "space bar" in the state space? It must be a state adjacent to each "letter" so that to provide the freedom a word to start and end with (almost) every "letter." An immediate outcome of this view is that the motion in the state space is to be orbital. Thus a necessary for the simplest semantics condition imposes certain structure of the state space: it must consist of at least 3 basins-of-attraction and to have accumulation point which is tangent to each of the basins. Then, the motion is indeed orbital so that each orbit is tangent to the accumulation point. What provides the crucial difference with a random sequence of letters though separated by "space bar" is that the bounded motion allows transitions only to adjacent basins. Thus, the choice of further letters is limited to those whose basins are adjacent.

Thus, the above structure of the state space provides two major differences with a Turing machine: the first one is that a hierarchically self-organized system spontaneously "says words" in response to the environment. Further, these words are not random sequences of letters, space bar included. What makes the sequence to have autonomous semantics is the restraint established over the possible succeeding of letters: for example "a" cannot follow "o." Thereby, the set of possible words is not the total set of all possible combinations of letters. Then, the major difference between the so organized primitive semantics and a Turing machine is that "hardware" becomes active: it selects a subset of semantic units (words in the case considered) so that to "express" itself. The advantage of this form of "response" is that the system stays stable. Moreover if a letter is missing in recording we can easily identify it because we know the admissible set of its predecessors and successors; we can even "foresee" the future state of the system on the grounds of the knowledge about the current state only. Thus, the property of being an "oracle" substantiates the major advantage of the semantic-like response: it provides its non-extensivity. The non-extensivity constitutes the major difference with the traditional computing. Indeed, the extensivity of the computing commences from the assertion that every algorithm is reducible to a number of arithmetic operations, executed by means of linear processes. Thus, the more complex is the algorithm the more operations

it involves and in turn, the more hardware elements (and/or time) are necessary to accomplish the task. Along with the extensivity, the traditional computing is non-autonomous: it execution is governed by an "external mind" who sets the algorithm and who comprehends the results.

The spontaneous limitation of the admissible neighbors for every letter is the first level of setting grammatical type rules in the organization of response. Yet, there is a very interesting parallel between grammatical rules, causality and the 'arrow of time'. Each time when a system works as "oracle" the corresponding transition from one basin-of-attraction to the only admissible one could be considered as a deterministic causal relation between the corresponding laws. An exclusive property of that deterministic relation is that it holds even in varying environment. In cases where the admissible basins are more than one, the random choice of one selection could be considered as "free will." Let me remind once again that the properties of being "oracle" and of the "free will" are properties of the self-organization of certain natural systems and thereby are not to be taken as exclusive properties of some superior human mind.

Further, the admissible orbits that substantiate words reveal long-range correlations among the letters in the corresponding words; we have already considered that these trajectories are non-Markovian. This opens the door to assigning "arrow of time" to each orbit. Its origin is to be associated with the sensitivity to permutations in the succession of letters. There is highly non-trivial interplay between the circularity and the linearity of the "arrow of time": it is locally linear, i.e. it is linear while writing a word; when finishing writing it turns circular; at the next level (writing a sentence) it is again linear, when putting a dot it becomes circular. It is worth noting that such interplay between circularity and linearity of the arrow of time is substantiated by the idea of associating the semantics with the natural processes. It should be stressed that neither philosophical tradition, not western nor eastern, assigns the property of the time to be simultaneously linear and circular. In the context of the modern inter-disciplinary science we could say that the time is not scale-invariant. It is more appropriate to say that time is a characteristic of the dynamics of the response and thus its properties depend on the local dynamics. Viewed from this perspective, time has the same origin and role as the metric in the state space: they both commence from the dynamics and are function of the hierarchical organization. The major role for the interplay between the circularity and linearity of the time is the non-extensivity of the hierarchical order.

Yet, we have not answered the question whether there are natural systems which have the above properties. These are all systems that are open to morphogenesis, starting with the simplest extended open reaction-diffusion systems which meet the condition of spatial heterogeneity. The spatial heterogeneity is viewed as necessary condition for the simplest hierarchical self-organization whose state space retains the desired properties. Further, more complex semantics is achieved through further hierarchical self-organization of networks whose self-organization constitutes their intra- and inter-level structure and functioning. It should be stressed that while the traditional networks are described predominantly in probabilistic way, here we put the stress on the compatibility and synchronization between their structure and functioning thereby approaching them from completely different perspective.

NON-EXTENSIVITY OF HIERARCHICAL STRUCTURING

So far we have considered the hierarchical structuring as the major implement for strengthening the response. This goal has been achieved by considering the role of the inter-level feedbacks. Indeed, through the appearance of each of them as a bounded stochastic noise, they provide a non-

ambiguous choice of units which in turn define the corresponding basins-of-attractions as intrinsic property of a system. This fact turns crucial for identification of the system. The identification of a system, i.e. its distinction from any other is one of the most fundamental questions about the behavior of the complex systems. On the one hand, the enormous variety of the complex systems precludes the possibility for any general rules for achieving both identification and diversity. On the other hand, the persistent coexistence of specific and universal properties suggests that there must be a general rule that produces the interplay between the identification and diversity along with the persistent separation to specific and universal properties. So far, our attention has been focused predominantly on establishing the universal properties and separation to universal and specific ones. Now we will present the most general implement for achieving diversity of the complex systems.

We start our considerations by putting the stress on functional diversity rather than on the structural one. The reason for this is that the achievement of structural and functional diversity is grounded on different principles. Thus, while the structural one is grounded on laws like energy and mass conservation in every elementary act, the functional diversity is grounded on algorithmic and semantic unreachability of one unit from another. The fundamental difference between mass/energy conservation and algorithmic/semantic unreachability commences from the fact that while the conservation laws require that the corresponding variables, mass, energy etc., retain metrics, the algorithmic/semantic unreachability implies metric-free way of defining the corresponding properties. Thereby, the functional properties are expressed in a fundamentally different way from the corresponding structural ones.

Beyond doubt the hierarchical structuring is a major implement for strengthening the response. Yet, the following consideration demonstrates that the strengthening of the response goes through its diversification: since the major implement for strengthening of the response is the self-organization which implements emerging of new properties, the resulting hierarchical organization turns out to be the desired implement for reaching diversity. Yet, now we face the dilemma: whether the obtained diversity is of a reductionist type, i.e. whether the laws at the higher levels are entirely dictated by the laws at lower levels. A positive answer to this question would put us back to the traditional reductionist philosophy of physics. Indeed, the latter reduces the complexity to separate levels of organizations (elementary particles, atoms and molecules, large systems, biological systems) each of which is considered at specific pre-determined external constraints so that the hierarchical order goes only bottom up i.e., from elementary particles to living organisms. These assumptions constitute its fundamental difference from the concept of boundedness which, on the contrary, sets hierarchical self-organization at non-specified external constraints so that its maintaining goes both bottom up and top down by means of loop-like inter-level feedbacks that serves as constraints imposed by one level to the nearest (both upper and lower) ones. Further, on the contrary to the reductionist approach which adopts the idea of dimension expansion as a tool for reaching diversity of properties, the diversification under boundedness happens through hierarchical super structuring in constraint dimensionality.

To the most surprise, our answer to the question whether the higher level properties commence for the lower level ones is neither positive nor negative. Actually we assert that the hierarchical organization has a two-fold comprehending: the one related to the structural properties operates in a reductionist way while the one related to the functional properties operates in a non-reductionist way. Thereby the non-trivial interplay between both of them brings about the non-extensivity of the hierarchical organization. Let us now consider in more details the assertion that the hierarchical organization retains both reductionists and non-reductionists properties: as we already assumed, the structural properties subject to relations that involve metrics, such as energy/mass conservation, are reductionist; on the contrary, the dynamical properties which does not involve metrics are non-reductionist. The first evidence for non-reductionist properties is justified by the appearance of a non-recursive component in the power spectrum of self-sustained oscillations. It is a metric-free property because it appears in the discrete band of the power spectrum as a non-recursive component whose role is to indicate a ban over resonances by bounded stochastic motion around the discrete band. Note, that this constitutes the major difference both with the traditional idea of resonance and with the idea the predictability of a stochastic motion. Indeed, the traditional "metric" notion of a resonance implies a specific quantitative relation between frequencies so that, on its fulfillment, the amplitude grows to infinity. The idea of traditional predictability of the future behavior of a stochastic sequence also implies specific metric properties: if the value of the sequence is this or that, in the next moment it would be so and so. Instead, the non-recursive component in the power spectrum indicates that the motion in the state space is permanently bounded and its boundedness is expressed by two components: the periodic motion which commences for self-sustained oscillations and stochastic motion which is characterized by a continuous band of the shape $1/f^{\alpha(f)}$ .The discrete band appears as an intra-basin invariant and so it characterizes the "homeostasis" while the shape of the continuous band is robust to the statistics of the fluctuations that constitutes the inter-level feedback. Moreover, as we proved in the previous section, both components of the motion retain the same metrics. To remind, as discussed in section *Metrics and* $1/f^{\alpha(f)}$ noise in Chapter 2, the metrics appears as a minimal condition for the continuity of the shape $1/f^{\alpha(f)}$ viewed as a single fit for the entire continuous band. In turn, the continuity of the shape $1/f^{\alpha(f)}$ provides its insensitivity to fluctuation statistics and its coarse-graining. Two major consequences come out from these results: (i) the law comprised by the corresponding basin-of-attraction meets the principle of relativity, i.e. it is independent from the auxiliary reference frame chosen for its observation; (ii) the knowledge about the law obtained from the power spectrum is robust to the small changes in the environment; it is also robust to the inevitable non-linear and non-homogeneous distortions in recording (measurement). Thus, the predictability is substantiated by providing constant accuracy for reoccurrence of the discrete band (which characterizes the homeostasis) in an ever-changing environment.

Further, the presence of a non-recursive component in the power spectrum implements its algorithmic unreachability from another law, i.e. from a law which is valid in another basin of attraction and is characterized by another power spectrum with another non-zero discrete band. Thus, the algorithmic unreachability provides diversity of laws, i.e. diversity of "identities." These properties of the power spectrum prove its relevance for associating each discrete band, non-recursive component included, as with an information symbol. It is worth noting that the property of being metric-free constitutes a uniform way of "comparing" the algorithmic properties of different information symbols besides their algorithmic unreachability; thus though being algorithmically unreachable, the physical distance between any two information symbols, viewed as involved and exchanged matter/energy is finite in metric sense.

Outlining, the non-extensive interplay between metric and metric-free properties allows obtaining of both the structural and the functional diversity so that while the structural diversity is of reductionist type, the functional one is not. Alongside the hierarchical structuring implemented by means of self-organization provides diversification of identities though acquiring them with algorithmically unique properties.

SEMANTIC UNIT AS AN ENGINE

So far we have considered the most general outcomes of the distinction between the structural and functional properties of the complex systems. We found out that the distinction goes via a remarkable separation of the properties to metric-dependent and metric-free. We have established that the metric-dependent properties are to be associated with the structure of a system while the metric-free are to be assigned to its functionality. The wisdom of this separation is that thereby the structural properties are governed by quantitative relations like mass and energy conservation laws whose major merit is that they provide reducibility of every structure to constituents; put it in other words it ensures reducibility of a molecule to atoms and at the same time constitute building of a piece of rock by a finite selection of molecules, etc. On the other hand, characterization of the functional properties in a metric-free way opens the door to algorithmic diversity of behavior: each functionality is unique and algorithmically unreachable from any other. Thus the separation of the structural and functional properties of the complex systems serves as grounds for implementing non-recursive computing whose major property is that while the information symbols (discrete bands in the corresponding power spectra) are algorithmically unreachable from one another, they are still physically reachable by means of involving and exchanging finite amount of energy/matter in the transition from one information symbol to another.

Even though the non-recursive computing is irreducible to a Turing machine it is still a non-autonomous process. Indeed, if not constrained, the admissible transitions cover the entire set of possible transitions. Thus, any sequence of information symbols could be produced. We would obtain just another type of a non-autonomous computing machine. Like the traditional Turing machine it neither "constructs" a meaningful sequence nor comprehends it in an autonomous way: it still needs external mind both for constructing a meaningful sequence and for its comprehending.

The factor that makes the non-recursive computing autonomous and fundamentally different from a Turing machine is the self-organization of the non-recursive computing in a semantic-like manner. In the section *Structure and Topology of the State Space* in the present Chapter we have established the minimal necessary condition for the simplest semantics, i.e. writing words, is the state space to comprise at least 3 basins-of-attraction and one accumulation point, i.e. a point tangent to all domains. Then, to each orbit which passes through the accumulation point a "word" is associated. Thus, the accumulation point plays the role of a "space bar," i.e. a natural separator of words. Further, the sequence of the letters in a 'word' is not arbitrary: each and every following letter is a random choice from the admissible to that given letter ones; yet, the set of the admissible letters is a subset of all possible ones and comprises only the nearest neighbors. The restriction over the admissible letters is brought about by the boundedness of the rates: the spontaneous motion in the state space is exerted by steps whose size is bounded by threshold set by the properties of the corresponding inter-level feedback. Thus, since the size of every basin-of-attraction is much larger than the largest step in the stochastic motion, it becomes obvious that the only admissible "letters" (basins-of-attraction) from any given one are its nearest neighbors. So, the spontaneous motion in the state space "writes" "words" which are *not* a random sequence of letters. Yet, in order to assign the notion of semantics to spontaneous "writing" of "words" we still miss implementation of the sensitivity to permutations of the letters in a word. It is a common knowledge that the sensitivity to permutations is one of the major generic properties of the semantics: any change of the order of the letters in a word, any change in the order of the words in a sentence etc. radically changes its meaning. Thus, we can distinguish without a moment of hesitation between the words "*dog*" and "*God*" for example. Moreover, we associate a specific meaning of each of these words which

is *irreducible* to the permutation of the constituting letters only.

Further, along with sensitivity to permutations we face another question: how to express in an algorithmically unreachable way the difference between a sequence of letters and the corresponding "word." Put it in other words how to distinguish between the information content of a word (the sequence of letters) and its meaning? The answer to this question is positive, i.e. there is a generic way of distinguishing the content of a word from its meaning. It goes through associating of the meaning of a semantic unit with an engine "built" on the orbit which represents the corresponding semantic unit. One of the central results in our book will be that any such engine is equivalent to a specific to it Carnot engine so that the efficiency of the original engine never exceeds that of the corresponding Carnot one. Thereby the functional irreversibility of a Carnot engine, i.e. its generic property to work as a pump in one direction and in the opposite direction to perform as a refrigerator, provides the target sensitivity to permutations. Further, the functional irreversibility of a Carnot engine provides algorithmic irreducibility of its performance to the specific processes that it exerts. In turn, the associated with a given Carnot engine "word" retains autonomous comprehending that is irreducible to the constituting "letters."

Further, the organization into cycles, viewed as a necessary condition for steady functioning of a structured network prompts suggestion about the uniformity of the hierarchical structure of semantics: we assert that at any level of hierarchical organization a semantic unit (e.g. "word") is constituted by a specific sequence of "letters" so that its semantic meaning to be represented by the specific performance of the corresponding engine. At the next hierarchical level the "sentence," comprised by several "words," represents a new syntactic unit which also must have an autonomous description irreducible to the sequence of "words" on the lower hierarchical level and so its semantic meaning is to be associated with the performance of another engine.

It should be stressed that the association of the meaning of a semantic unit with the corresponding engine substantiates the algorithmic irreducibility of the meaning of the semantic unit to the sequence of units that constitute it. Indeed, the characteristics of any engine are the work that it produces and its efficiency. Yet, even though each of them is specific for the corresponding cycle, neither of them is reducible to the specific characteristics of the cycle.

It is worth noting, that the organization of the semantics in a non-extensive hierarchy of engines, each of which performs in a specific way, suggests their "orchestrating" so that the system as a whole to stay stable arbitrary long-time. This implies presence of long-range correlations among the functionality of engines both inside each level and among levels. In turn, the long-range correlations operate so that to prevent arbitrary large accumulation of matter/energy anywhere in the system. At the same time the "orchestrating" implements the long range correlations typical for any piece of semantics. Thus, we again encounter the miraculous interplay between the issue about the stability and the intelligence. And the conclusion drawn from the present considerations is that the more intelligent the system is, i.e. the more sophisticated its hierarchical organization is, the more stable it is, i.e. it remains stable for the larger variety of stimuli.

Outlining, the non-extensive hierarchical self-organization of the response in a semantic-like manner justifies its fundamental difference from a Turing machine: it turns out that the self-organized semantics is irreducible to a random sequence of units and could be "comprehended" non-ambiguously by a similarly organized object/subject. Thus, the shared functional similarity serves as grounds for shared comprehending of the semantics viewed as diversified complex response of hierarchically self-organized object/subject. It is worth noting that the semantics of the response

retains the major properties of its traditional understanding, namely its non-extensive structuring and sensitivity to permutations at every level of its hierarchy.

It is worth highlighting once again the major difference from a Turing machine where hardware silently "follows" the algorithmic properties of the software. Then: (i) the greatest value of a Turing machine is that it can execute any recursive algorithm; however, this comes at the expense of non-autonomous comprehending of the obtained output. On the contrary, the ability of the introduced by us non-recursive, semantic-like computing to execute any algorithm is constrained by the boundedness which precludes transitions which exceed thresholds; at the same time, it substantiates the fundamental value of that approach – autonomous comprehending. In turn this makes the Turing machine and the semantics to appear rather as counterparts than as opponents.

The story about the association of the semantic meaning with an engine has just begun - there are a lot of other important matters to discuss. One of the most exciting ones is the issue whether this idea supports the Second Law. The problem is that this idea encounters fundamental difficulties with the most widespread formulation of the Second Law, namely the one that asserts: the equilibrium is the state of maximum entropy. We start revealing this controversy by pointing out that the traditional notion of entropy encounters fundamental difficulties when applied to structured systems since any order and/or structure implies by definition that the corresponding state is *not* the state of maximum entropy. Alongside, the replacement of the idea of equilibrium with that of homeostasis and the assumption that neither basin of attraction is global attractor substantially contributes to this controversy. On the other hand, the idea of implementation of semantic meaning through a specific "engine" implies that semantics is "hard work" supplied by the environment. Thus it is important to quantify the relation between the "work" produced by the "engine" and the energy/

matter involved in the corresponding process. So, we face the question whether our idea yields a "perpetuum mobile" and if not, how and why it happens; is the ban over "perpetuum mobile" a general result and if so how the concept of boundedness modifies the idea of the Second Law. These issues are subject matter of the part B of our book.

CONCLUSION

The ubiquitous coexistence of specific and universal properties of the complex systems poses the question whether there exists a general protocol that governs its persistence. The enormous difficulty in its resolving lies in the supposition: if exists such protocol should not only substantiate the persistence of the separability of the response to a specific and universal part but along with it must open the door to diversification of the specific part. Our answer to the question is positive, i.e. we assert that such general protocol do exists and it consists of the hierarchical order that is the major implement for strengthening the response through its specification at different hierarchical levels. Thus, on the one hand, the hierarchical structuring serves as an implement for enhancing the stability of a complex system. On the other hand, the specification of the response viewed as response to different external stimuli at different hierarchical levels opens the door to the diversity of the response.

The major move ahead commences from adoption of the idea that the major implement of the hierarchical self-organization of the response is the self-organization. The idea of self-organization is one of the most fascinating concepts of modern interdisciplinary science. It assumes that when tuned to certain values of control parameters, a system displays "emergent" properties, i.e. properties that are not "encoded" in its micro-dynamics. A paradigmatic example of emergent behavior is the so-called self-sustained oscillations which emerge from a non-driven autonomous homogeneous system. Our implementation of the idea of self-organization goes via its modification so that to transform it into an operational protocol whose aim is to involve the inter-level feedbacks. Involvement of the inter-level feedbacks is necessary in order to describe the links among the levels which keep the system together. The greatest advantage of this modification is that it brings about non-ambiguous choice of the units (they are set by the use of intrinsic for the system scales only) which in turn allows robust to small environmental fluctuations determination of the corresponding basins-of-attraction. Alongside, the self-organization preserves its original property to give rise to emergent phenomena. Further, the incorporation of inter-level feedbacks into the equation-of-state provides two exclusive generic properties of the power spectrum of its solution: (i) the power spectrum is additively decomposed to a specific discrete and a universal continuous band so that the shape of the continuous band is insensitive to the statistics of the fluctuations; (ii) the discrete band comprises a non-recursive component. Thus, our first goal is achieved: the separation of the response to a specific and universal part turns out to be an exclusive generic property of the equation-of-state for the self-organization. It should be stressed that it appears as a generic property exclusive for the fluctuations under the mild constraint of boundedness. It is worth noting that the only constraint over the fluctuations is that of the boundedness over their size and rate; the fluctuations are free from any pre-determination like obeying certain distribution, linearity of the response etc.

Therefore, the first great success of the present modification of the self-organization is that it substantiates the persistency and ubiquity of the separation of the response to a specific and universal part. But it gives more than that: it justifies the certainty of our knowledge about this separation. Indeed, since the fluctuations which participate in the inter-level feedback are not subject to pre-determination, their recording is subject to inevitable non-linear and non-homogeneous distortion. An exclusive property of the power spectrum is that the shape of the continuous band is insensitive to these distortions and thus provides constant accuracy for the reoccurrence of the specific discrete one. Thereby, the identification of the system expressed through the discrete band is kept intact in an ever-changed environment.

Yet, the self-organization is more than that: the generic structure and the topology of the state space of the simplest open flow systems which are subject to morphogenesis appears as a natural implement for organization of the response in a semantic-like manner. Indeed, since each basin-of-attraction is characterized by a specific invariant (discrete band in the power spectrum), the latter could serve as a "letter" in the corresponding "alphabet." Further, the accumulation point, tangent to every basin-of-attraction, serves as a "space bar." Therefore, the spontaneous motion in the state space, i.e. the response of a system, "writes" words which are *not* random sequences. This assertion is justified by the fact that the coarse-grained state space motion, i.e. the inter-basin motion, goes only via adjacent basins; this happens because of the boundedness of the rates: the largest fluctuation is by definition much smaller than the size of the corresponding basin-of-attraction. Thereby, the "words" are not random sequences of letters. An exclusive property of thus organized semantics is its spontaneous creation and its autonomous comprehending by a similarly organized object/subject.

The self-organization not only substantiates the semantic-like organization of the response but provides a general approach to its diversification: indeed, it goes through using the output of one functional unit as the input for another; for example, the simplest form of emergent phenomenon, self-sustained oscillations, could be used for driving another unit; in turn the non-linear interaction between driving and emergent phenomena at that unit would result into something new, emergent, which again could be used as driving force etc.

Thereby, it is possible to organize a "network" consisting of "chains" of mutual driving. Thus, the self-organization ensures the desired structural diversity of complex systems. It provides not only the structural diversity but the functional one as well: the ubiquitous presence of a non-recursive component in the power spectrum makes different functional units algorithmically unreachable one from another which in turn provides their diversity.

The major move ahead in providing diversity comes out from the association of a semantic unit with an engine built on the corresponding orbit. Thus, the meaning of a semantic unit is not reducible to the sequence of letters by which it is comprised. Moreover, the functional irreversibility of the engines provides the sensitivity to permutations of the letters in a word. It is worth noting, that the organization of the semantics in a non-extensive hierarchy of engines, each of which performs in a specific way, suggests their "orchestrating" so that the system as a whole to stay stable for indiscriminately long-time. This implies presence of long-range correlations among the functionality of engines both inside each level and among levels. In turn, the long-range correlations operate so that to prevent arbitrarily large accumulation of matter/energy anywhere in the system. At the same time the "orchestrating" implements the long range correlations typical for any piece of semantics. Thus, we encounter a surprising interplay between the issue about the stability and issue about the intelligence. And the conclusion drawn from the present Chapter is that the more intelligent the system is, i.e. the more sophisticated its hierarchical organization is, the more stable it is, i.e. it remains stable for the larger variety of stimuli.

REFERENCES

Feller, W. (1970). *An introduction to probability theory and its applications*. New York, NY: John Wiley & Sons. doi:10.1063/1.3062516

Chapter 5
Invariant Measure:
Power Laws

ABSTRACT

Bounded randomness of mass/energy exchange rates neither presuppose nor selects any specific time scale, thresholds of stability included. Nonetheless, the boundedness of the rates sets certain non-physical correlations among successive increments and thus justifies formation of "sub-walks" on the finest scale. Further, the "U-turns" at the thresholds of stability set certain correlations on the biggest possible scale of a relevant variable. The major question now is how the balance between the universal correlations, set by the "U-turns," and those of the specific "sub-walks," set by the bounded randomness, shapes the structure of a BIS that represents the evolutionary pattern of a relevant variable. It is proven that this issue is inherently related to another universal property of complex systems behavior that is power law distributions. It is demonstrated that power law distributions acquire novel understanding in the setting of boundedness: they appear as universal criterion for hierarchical structuring implemented under boundedness.

INTRODUCTION

The major aim of the previous chapter has been to demonstrate that the self-organization of a complex system in a multi-level hierarchical manner provides strengthening of the response by means of its diversification: different levels respond to different stimuli. The connectivity of a system implies that the hierarchy of levels is "linked" by inter-level feedbacks each of which operates in both directions, i.e. bottom up and top down. In Chapter 3 we considered the basic

DOI: 10.4018/978-1-4666-2202-9.ch005

inter-level feedback, namely the one that operates at quantum level and makes atoms and molecules to self-organize in a" flow" with certain concentration. Moreover, we established that at the quantum level the concentration controls the local accumulation of energy/matter so that to intensify the dispersion of any local exceeding of matter/energy accumulation; in turn, on the next hierarchical level the "controlling mechanism" appears as boundedness of the local fluctuations of the concentration. Thus the inter-level feedback operates in both directions. The high non-triviality of the matter is 3-fold:

1. It starts with imposing boundedness at the quantum level by means of assuming boundedness over local accumulation of matter/energy as leading property. It has been proven that this could not be achieved by means of imposing boundedness over the velocities alone. To the most surprise it turns out that this boundedness not only does not prevent arbitrary accumulation of matter/energy but, on the contrary, it produces amplification of the local fluctuations. The successful approach turns out to be the consideration of non-unitary interactions in the setting of the concept of boundedness so that their generic property, dissipativeness, to be used for dispersing any extra-accumulated matter/energy.

2. Thus at the next hierarchical level the inter-level feedback operates as a source of inevitable fluctuations of the concentrations whose generic property is their boundedness. What sets their appearance as an inter-level feedback but not as a bounded environment is their participation in the choice of units. It is worth reminding that the choice of units is of primary importance for non-ambiguous setting of the structure of the state space when the evolution is described by non-linear equations of any type. Then, as considered in section *Metrics in the State Space* of Chapter 4 the inter-level feedback participates with units determined at the lower level while a bounded environment sets units external for the system. Thus, in the latter case a system does not retain robustness to an ever-changing environment and thus it would not be time-translational invariant.

3. The above properties of the basic inter-level feedback along with the property of bringing about emergent phenomena make self-organization the appropriate generic implement for diversification of the response through further hierarchical super-structuring. Therefore the diversity of properties is reached by diversification of the hierarchical structuring. It should be stressed that this setting makes our approach fundamentally different from the traditional reductionist philosophy of physics. Indeed, the latter reduces the complexity to separate levels of organizations (elementary particles, atoms and molecules, large systems, biological systems) each of which is considered at specific pre-determined external constraints so that the hierarchical order goes only bottom up i.e., from elementary particles to living organisms. These assumptions constitute its fundamental difference from the concept of boundedness which, on the contrary, sets hierarchical self-organization at non-specified external constraints so that its maintaining goes both bottom up and top down by means of loop-like inter-level feedbacks that serve as constraints imposed by one level to the nearest (both upper and lower) ones. Further, contrary to the reductionist approach which adopts the idea of dimension expansion as a tool for reaching diversity of properties, the diversification under boundednes happens through hierarchical super structuring in constraint dimensionality.

A key assumption we have made is the consideration that hierarchical structuring is a common property for both intelligent and non-intelligent systems. Yet, the intelligent-like systems retain additional property that their response is executed in a semantic-like manner. The major property of the semantic-like response is that it is active, i.e. it is not reducible to a simple general relation with the impact. What is very important is that the "active" response is not exceptional but it is rather typical for very large amount of systems: we have established that the simplest semantics is executed under the mild condition for partitioning of the state space in 3 basins-of-attraction and one accumulation point. It turns out that every open spatially heterogeneous system fits this condition. Thus, indeed, the semantic-like response is rather typical than exceptional property.

The property of the response of being active, i.e. not to be reducible to a single relation with the local environmental impact poses the fundamental question whether it is ever possible that the behavior of such system is subject to time translational invariance. The issue seems highly controversial since on the one hand we have demonstrated that the power spectrum of the solution of the equations that describe self-organization are additively decomposable to a discrete and a continuous band of shape $1/f^{\alpha(f)}$ type. The importance of this decomposition lies in providing constant accuracy of the discrete band reoccurrence in an ever changing environment. Further, being an intra-basin invariant, the discrete band appears as an appropriate candidate for a "semantic unit" whose time-translational invariance seems inherent. On the other hand, the presence of an accumulation point with a special property of being tangent to every basin-of-attraction poses the question whether it signals out a specific time scale.

Further in this line of reasoning comes the question whether permutation sensitivity of the semantics sets certain "ordering" of the time scales.

Last but not least there comes the question whether U-turns at the threshold of stability do not involve any physical process behind their execution. Thus, we face the very important questions: (i) whether a time series which comprises semantics has the properties of a BIS considered in Chapter 1; (ii) how the execution of U-turns happens so that not to involve a specific physical process behind it. The positive answer to these questions will provide time translational invariance of a semantic-like response. Put in other words it will provide that the semantics neither involves nor signals out any special time point. In turn, the time translational invariance of a semantic sequence ensures its reoccurrence with constant accuracy even if the environment does not re-occur exactly the same. Thus, the time translational invariance of the semantics is classified as a property that renders its universality and autonomy. This constitutes the major goal of the present Chapter: to demonstrate that a semantic-like response obeys time-translational invariance. It will be proven that this issue is inherently related to another universal property of complex systems behavior: power law distributions. It will be demonstrated that power law distributions acquire novel understanding in the setting of boundedness: they appear as universal criterion for hierarchical structuring implemented under boundedness.

BACKGROUND

The above considerations give rise to the question about the coexistence of diversity and universality. One the one hand there is a remarkable separation of the properties of a complex system into universal and specific parts. An exclusive property of that separation is that it happens at every level of the hierarchical self-organization. On the other hand, we assert that the diversity of specific properties is achieved by means of the developed by us in the previous Chapter modification of the idea of self-organization. The most advanced step in this new understanding has been the introduction of the idea that, at very mild conditions, the motion in the state space is organized in a semantic-like manner. Thus we face the question whether the trajectory of a semantics-like response retains the properties of BIS. An affirmative answer to this question would justify the remarkable universality and ubiquity of the separation to specific and universal parts of the properties of the complex systems. Moreover, by sharing that separation both intelligent and non-intelligent systems would give a credible argument in favor of assertion that the intelligence is an exclusive property of multi-level hierarchical self-organization.

The major problem we face is about the role of the accumulation point: is it a point of special quality and if so does it have a specific contribution to the power spectrum or to any other characteristics of the corresponding time series? The answer to the first question is both affirmative and negative: it is affirmative because on the one hand it is a point which is tangent to all basins-of-attraction and thus it is a special one; and since the basins-of-attraction are fixed, consequently its position in the state space is also fixed. On the other hand, since the motion in a bounded attractor is recursive and thus orbital in every point, the accumulation point is just an ordinary point in the state space which shares the property that the corresponding orbits are tangent to it. Thereby, the difference between a "semantic" trajectory and an ordinary trajectory is to be rather associated with the properties of the power spectrum of the basins-of-attraction through which it passes than with the structure and the properties of the state space. Indeed, the general property of any motion in the state space is that it is orbital and that each trajectory is a dense transitive set of orbits. Put in other words, the latter statement implies that a trajectory is any sequence of orbits; the size of an orbit is limited only by the corresponding distance to the boundaries of the attractor but it does not depend on its position in the state space.

Thus we can conclude that the "specialty" of accumulation point does not affect the property of a semantic-like trajectory to be a BIS.

The next question we face is whether the characteristics of orbits are specific; if so, it is to be expected their specific contribution to the power spectrum. This question is very important in view of the execution of the U-turns. To remind, the core of our concept of boundedness lies in the assertion that the boundedness of mass/energy exchange rates neither presupposes nor selects any specific time scale, thresholds of stability included. Nonetheless, the boundedness of the rates sets certain correlations among successive increments and thus justifies formation of "sub-walks" on the finest scale. Further, the "U-turns" at the thresholds of stability set certain correlations on the biggest possible scale of a corresponding variable. The major question now is how the balance between the universal correlations, set by the "U-turns," and those of the specific "sub-walks," set by the "bounded randomness," shapes the structure of a BIS that represents the evolutionary pattern of the variable. It is worth considering in details the origin of the specific "sub walks": it is to be associated with the non-trivial interplay between the "deterministic" and stochastic" terms in the evolutionary equations of the self-organization defined in the previous Chapter. To remind, in Chapter 4 we introduce the following general form of evolutionary equations (equation-of-state):

$$\frac{d\vec{X}}{dt} = \vec{\alpha} \bullet \hat{A}\left(\vec{X}\right) - \vec{\beta} \bullet \hat{R}\left(\vec{X}\right) + \vec{\mu}_t\left(\vec{X}\right) \quad (4.1)$$

where $\vec{X}$ is the vector of the concentrations; $\vec{\alpha}$ and $\vec{\beta}$ are the control parameters; $\hat{A}\left(\vec{X}\right)$ and $\hat{R}\left(\vec{X}\right)$ are the mass/energy exchanged in the processes taken at that hierarchical level and $\vec{\mu}_t\left(\vec{X}\right)$ are the fluctuations of the corresponding *concentrations induced by the non-unitary interactions (inter-level feedback).*

Next we have proved that these equations are equivalent to the following ones:

$$\frac{d\vec{X}}{dt} = \vec{\alpha}_t^{eff} \bullet \hat{A}\left(\vec{X}\right) - \vec{\beta}_t^{eff} \bullet \hat{R}\left(\vec{X}\right) \qquad (4.5)$$

where $\vec{\alpha}_t^{eff}$ and $\vec{\beta}_t^{eff}$ are effective control parameters chosen so that the instant solution of Equation (4.5) to match exactly the corresponding instant solution of Equation (4.1). Thus, at every moment the solution of Equation (4.5) deviates from the solution of Equation (4.4). The generic property of these deviations is their boundedness: to each point in the control parameter space is assigned certain threshold over the deviations which cannot be exceeded. Thereby the deviations rather appear as a random choice of a single selection of a bounded multi-valued function. The selections of that multi-valued function determine the set of admissible transitions from any given state.

An exclusive generic property of the solution of Equation (4.1) is that its power spectrum is additively decomposable to universal continuous band of shape $1/f^{\alpha(f)}$ a specific discrete band which comes from the equation:

$$\frac{d\vec{X}}{dt} = \vec{\alpha} \bullet \hat{A}\left(\vec{X}\right) - \vec{\beta} \bullet \hat{R}\left(\vec{X}\right) \qquad (4.4)$$

which determines the "homeostatic" part of the response. An exclusive property of Equation (4.4) is that the power spectrum of its solution is an intra-basin invariant. Then, the choice of the control parameters in Equation (4.1) inside each basin-of-attraction is ignorant with respect to the equivalence between Equation (4.1) and Equation (4.5). This justifies the assertion that all points in the state space are equivalent with regard to be considered as "sources" for an orbit; moreover,

since the choice of the control parameters in Equation (4.1) is free inside a basin-of-attraction, the size of the corresponding orbit is accordingly limited only by that choice but is independent from the position of the orbit in the state space. So, indeed, the size of an orbit is independent from its position in the state space and is limited only by the distance to the boundary of the state space.

Summarizing, one concludes that the sub-walks on finest scale are specific for the system. Yet, the fundamental role of the correlations set by the boundedness on each and every scale suggests that the specific "sub-walks" gradually "coarsen" so that to acquire universal parameter-free shape which makes them not to signal out any specific component in the power spectrum associated with it. The multi-valuedness of the stochastic terms in Equation (4.1) renders existence of more than one admissible state to any given one. This is matched in Equation (4.1) through the random choice of one selection from all available at every instant. Even though only one transition takes place, the set of all available ones constitutes the current range of the admissible transitions. The outcome is that the trajectory appears as a kind of a fractal Brownian walk. In addition, the stochastisity induced by the random choice of one selection among all available breaks any possible long-range periodicity and thus provides a finite size of any correlations along a trajectory. So, the trajectories in the state space are indeed BIS. We will prove in the next section that the shape of the sub-walks gradually acquires a parameter-free form which in turn serves as grounds for a mechanism of stretching and folding for the U-turns whose major property is that it allows automatic execution of U-turns without involving physical processes behind their execution. It is worth noting its fundamental difference from the traditional approach of imposing tangent boundary conditions: the latter implies involving of specific physical processes behind their execution.

Thus the difference between a "semantic" "trajectory" and any other admissible trajectory in the state space consists in the following: a "semantic" trajectory is an admissible trajectory which passes through the accumulation point along with "visiting" at least 3 basins-of-attraction. So, the "semantic" trajectories share the general property of all admissible trajectories to be a BIS. So, it is to be expected that it shares all generic for a BIS properties.

The present considerations suggest that there is another general characteristic of a BIS: invariant measure. The importance of such measure lies in its anticipated universality and ubiquity: if exists, it is to be equally available for both intelligent and non-intelligent complex systems and it is to be available at any level of hierarchical organization. It should be stressed that empirically such measure does exist and it is well known as power law distributions: it turns out that all complex systems are subject to this distribution. In the Preface to the present book we considered as an example of a power law distribution the Zipf's law. We will prove later in the present Chapter that universal measure assigned to the bounded motion in the state space successfully reproduces the major characteristics of the power laws and at the same time releases them from their major flaw: their clash with the time translational invariance.

Yet, it should be explicitly pointed out the difference between thus proposed invariant measure and the traditional probabilistic approach. This will be done in the section *Power Laws under Boundedness* of the present Chapter. Moreover, it will be demonstrated that power law distributions attain novel understanding: they appear as universal criterion for hierarchical structuring under boundedness.

STATE SPACE TRAJECTORIES: GENERAL CHARACTERIZATION

The general assumption we made about the motion in the state space is that it is orbital and such that the characteristics of the orbits are parameter free; this assumption is viewed as necessary for

providing uniform contribution of all time scales to the power spectrum. Thus we assume that there exists a scale a above which all scales up to the thresholds of stability contribute uniformly to the creation of orbits. Then, it is to be expected that the coarse-grained structure of the attractor exhibits universal properties insensitive to the details of the statistics of the stochastic terms in Equation (4.1); hereafter we refer the latter as transition rate statistics. Furthermore, it is evident that these properties must be similar in every direction of the attractor. The boundedness at scales beyond a will be referred hereafter as incremental boundedness. So, we will study the one-dimensional sequence produced by the projection of a trajectory onto any direction. We recall that every such sequence belongs to the class of BIS subject to incremental boundedness defined above.

In the present section we will study those universal properties shared by every trajectory that define the coarse-grained structure of the attractor. Bearing in mind that the state space trajectories are BIS, we will demonstrate that the coarse-grained over a structure of each trajectory is a succession of well separated from one another successive excursions. To remind, an excursion is a trajectory of walk originating at a given point at moment t and returning to it for the first time at the moment $t + \Delta$. The characteristics of each excursion are amplitude, duration and embedding interval. The latter is a property introduced by the boundedness and has no analog for the unbounded sequences. It implies that each excursion is loaded in a larger interval whose duration is interrelated with the duration of the excursion itself. The major role of the embedding is that it does not allow overlapping of the successive excursions and thus prevents growing of the excursion amplitude to arbitrary size. It results in permanent preserving of both static and the dynamical boundedness.

The incremental boundedness renders that each excursion has certain duration interrelated with its amplitude. By the use of this relation, we will prove in the section *Relation $A \leftrightarrow \Delta$. Sym-*

metric Random Walk as Global Attractor for the Fractal Brownian Motion. Finite Velocity that the velocity of the motion in the state space is always finite. The dilemma whether this velocity is finite or arbitrary is straightforwardly related to the issue of the dynamical boundedness. The fact is that an arbitrary velocity implies involving unlimited amount of energy/matter in a transition and its spreading through space and time with arbitrary velocity. On the contrary, bounded velocity implies involvement of a limited amount of energy/matter in each transition and its spreading through space and time with finite velocity.

In the next subsection we will prove that every coarse-grained trajectory appears as a sequence of separated by non-zero intervals successive excursions. It should be stressed that this structure is a result of the boundedness alone and does not depend on the statistics of the original trajectory.

Each excursion is characterized by its amplitude, duration and embedding interval as sketched in Figure 1.

The separation of the successive excursions means embedding of each of them into a specific to it larger interval so that no other excursions

Figure 1. Characteristics of the excursions

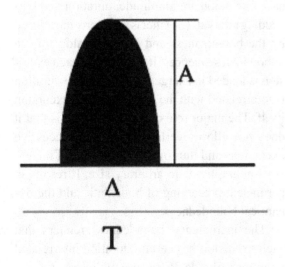

can be found in that interval. Below we will demonstrate that the duration of an "embedding" interval is a multi-valued function whose properties are strongly related to the duration of the embedded excursion Δ itself: the range and the values of the selections are set on Δ; the realization of any embedding interval is always associated with the realization of the corresponding excursion. Since the duration of each embedding interval is a multi-valued function, its successive performances permanently introduce stochasticity through the random choice of one selection among all available. Thus, the multi-valuedness of the embedding induces stochastisity that breaks any possible long-range periodicity (i.e. long-range memory) and helps the excursion sequence to preserve the properties established in Chapter 1 on coarse-grained scale.

The relation $\Delta \leftrightarrow A$ is set on "blob" structure of the walk that produces the coarse-grained trajectory, namely: because of the finite memory size, on coarse-grained scale any fractal Brownian walk can be considered as symmetric random walk of "blobs" created by subwalks whose size counterparts the memory size.

Our first task is to work out explicitly the relations $T \leftrightarrow \Delta \leftrightarrow A$, i.e. the relations between the duration of the embedding time interval T, respectively the duration Δ and the amplitude of the corresponding excursion A.

Embedding Time Interval

Relation $T \leftrightarrow \Delta$

The major role of the "embedding" is that it does not allow overlapping of the successive excursions and therefore prevents growing of the excursion amplitude to an arbitrary size. So, the "embedding" permanently "holds" the trajectories bounded.

The present task is to work out explicitly the relation between the duration of the embedding intervals and the duration of the corresponding excursions. That relation is based on the notion of excursion: trajectory of walk that originates at a given point at time t and returns to it for the first time at time $t + \Delta$. Therefore, the probability for excursion of duration Δ is determined by the integral probability for all pair of points to be separated by distance smaller than Δ. Taking into account that the probability that any two points separated by time interval η have the same value is given by the autocorrelation function $G(\eta, T)$, the probability that an excursion of duration Δ happens in an interval T reads:

$$P(\Delta, T) = \frac{1}{T} \int_0^\Delta G(\eta, T) d\eta = P(\Delta, T) = \int_0^{\Delta/T} \left(1 - x^{\nu(x)}\right) dx \tag{5.1}$$

We have already met the function $P(\Delta, T)$! Remember that in Chapter 1 it was the function $P(u)$ that gives the probability for a pair of points to participate to the stretching and folding. The different role in which it appears now comes from the difference in the definition of a BIS. In section *Autocorrelation function* in Chapter 1 we have considered the original BIS, whilst now we consider its coarse-grained counterpart. Therefore, the meaning of $P(\Delta, T)$ becomes different, namely Δ and T are associated with a single excursion at the coarse-grained counterpart, whilst at the original BIS they are associated with the long-range correlations. Nonetheless, the parity between short-range statistics of a single effective excursion in the coarse-grained counterpart and the universality of the long-range correlations in the original BIS renders parity between $P(\Delta, T)$ and $P(u)$.

The $P(\Delta, T)$ dependence only on the ratio Δ/T in (5.1) verifies the assumption that every excursion of duration Δ is "embedded" in an interval of duration T so that no other excursion happens in that interval.

The next step is to work out the shape of $P(\Delta, T)$. Its role is crucial for the behavior of an excursion sequence. To elucidate this point let us consider the following extreme cases:

1. $P(\Delta, T)$ is a sharp single-peaked function. Then it ensures single value of the most probable ratio $\frac{\Delta}{T}$. So, when the trajectory involves identical excursions, their appearance manifests rather periodic behavior which however, inevitably introduces long-range correlations;

2. $P(\Delta, T)$ has gently sloping maximum. Then, the relation between Δ and T behaves as multi-value function: range of nearly equiprobable but different values of T corresponds to the same Δ. Consequently, identical excursions are embedded in time intervals whose durations are randomly chosen among all equiprobable that correspond to their duration. In turn, the variability of the embedding time intervals induces stochastisity that breaks any long-range correlations along the trajectory.

The establishing of the shape of $P(\Delta, T)$ requires explicit knowledge about the shape of $\nu\left(\frac{\Delta}{T}\right)$. It is worked out on the grounds of the assumption that all time scales contribute uniformly to the properties of the BIS. Let me recall that we already have established in section *Autocorrelation function* in Chapter 1 that this requirement sets $\nu\left(\frac{\Delta}{T}\right)$ to be a linear function of its argument:

$$\nu\left(\frac{\Delta}{T}\right) = \left(1 - \frac{\Delta}{T}\right). \tag{5.2}$$

The plot of $P\left(\Delta, T\right)$ with the above shape of $\nu\left(\dfrac{\Delta}{T}\right)$ is identical to $P\left(u\right)$ presented in Chapter 1, Figure 1. It has a gently sloping maximum: indeed, the values of $P\left(\Delta, T\right)$ in the range $\dfrac{\Delta}{T} \in \left[0.25, 0.4\right]$ vary by less than 7%. Outside this range $P\left(\Delta, T\right)$ decays sharply. Thus, though $P\left(\Delta, T\right)$ is single-valued function, it provides nearly multi-valued relation between the most probable values of Δ and T, namely: certain range of nearly equiprobable values of T is associated with each Δ. In the course of the time the multi-valued relation is exerted as random choice of the duration of the "embedding" intervals. Thus, the execution of the multi-valuedness prevents the formation of any long-range correlations even when the sequence comprises identical excursions. In turn, the lack of long-range correlations puts a ban over appearance of extra-line(s) in the power spectra and thus ensures the time-translational invariance of its discrete band.

RELATION $A \leftrightarrow \Delta$

Symmetric Random Walk as Global Attractor for the Fractal Brownian Motion

Finite Velocity

The incremental boundedness gives rise to the anticipation that there is a general relation between the amplitude of each excursion and its duration. The relation amplitude $\leftrightarrow$ duration determines not only a property of an excursion but the "velocity" of the "motion" on that excursion as well. In turn, the latter sets the velocity of the motion in the state space. The question we pose now is whether the velocity of that motion is bounded?

It has already been established that each state space trajectory can be considered as a fractal Brownian walk. The latter provides the following general relation between the amplitude A and the duration Δ of an excursion, namely: $\sqrt{\left\langle A^2 \right\rangle} \propto \Delta^{\beta(\Delta)}$, where $\beta\left(\Delta\right)$ is set on the particularity of the transition rate statistics; the averaging is over the sample realizations. The dependence of β on Δ comes from the interplay of the finite radius of the correlations a and the amplitude of the excursion itself that is limited only by the thresholds of stability.

Because of the finite memory size, on coarse-grained scale any fractal Brownian walk can be considered as a symmetric random walk of "blobs" created by sub-walks whose size is counterpart of the memory size. The finite size of the memory renders the blob creating sub-walks to have finite length m; the particularities of the transition rate statistics determines the exponent ρ so that that the $\sqrt{m.s.d.}$ (where $m.s.d.$ stands for mean square deviation) of the blob creating sub-walks equals m^{ρ}. Then, the large excursions are approximated by symmetric random walk with constant step equal to the blob size. Thus, the dependence of any large scale excursions on its duration reads:

$$\sqrt{\left\langle A^2 \right\rangle} \propto N^{0.5} m^{\rho} \qquad (5.3)$$

where N is the number of the blobs.

It is obvious that when $N \gg m$ the dependence tends to:

$$\sqrt{\left\langle A^2 \right\rangle} \propto N^{0.5} a \qquad (5.4)$$

where a is considered constant independent of N. So, the symmetric random walk with constant step appears as global attractor for any fractal Brownian motion regardless to whether it is super- or sub-diffusional. Note, however, that this is valid only when the fractal Brownian walk is subject to boundedness and that it does not hold for arbitrary walks!

Let us now focus our attention on the limitations that dynamical boundedness imposes on the relation between the size and the duration of each excursion. The limitation commences from the assumption that a finite rate of development of each and every excursion requires one-to-one correspondence between them, i.e. each excursion of a finite size must have finite duration so that the rate of development to stay bounded in the prescribed range. This is automatically provided only for non-zero yet finite values of the exponent ρ. To make it clear, suppose $\rho = 0$. It makes the blob size insensitive to the short-range statistics which contradicts the major assumption that the blobs are fractal Brownian walks set on the short range statistics. Let us now suppose the other extreme: $\rho \approx \infty$; it immediately turns the blob size to infinity regardless to the details of the transition rate statistics. Note that non-zero yet finite values of ρ ensure non-zero but finite value of $\beta(\Delta)$. The velocity of the motion in the state space reads:

$$\frac{dA}{dt} = \frac{d\left(\Delta^{\beta(\Delta)}\right)}{d\Delta}\frac{d\Delta}{dt} \propto \beta(\Delta)\Delta^{\beta(\Delta)-1}\rho m^{\rho-1}$$

(5.5)

Obviously, every finite and non-zero combination $\left(\beta(\Delta), \rho\right)$ provides not only finite velocity but sets it non-zero as well. Thus, the motion in the state space is permanent and executed with finite velocity. Let us recall that this result is in sharp contrast with the traditional statistical mechanics where the velocity is arbitrary. The di-

lemma whether the velocity is finite or arbitrary is straightforwardly related with the issue of the dynamical boundedness. An arbitrary velocity implies involving unlimited amount of energy/matter in a transition and its spreading through space and time with arbitrary velocity. On the contrary, bounded velocity implies involvement of a limited amount of energy/matter in each transition and its spreading through space and time with finite velocity.

An important outcome of the above considerations is that the finite size of the "blobs" ensures uniform convergence of the average to the expectation value of the original trajectory. The distinctive property of any fractal Brownian walk is that any exponent $\rho \neq 0.5$ arises from arbitrary correlation between the current increment ς_i and the corresponding step τ_i. Then the average $\bar{A}$ reads:

$$\bar{A} = \sum_{i=1}^{N}\varsigma_i\left(\tau_i\right)\tau_i = \sum_{i=1}^{N}(-1)^{\gamma_i}\tau_i^{\rho_i}$$

(5.6)

and correspondingly the *m.s.d.*:

$$\left\langle A^2 \right\rangle \propto \left\langle \sum_{i=1}^{N}\left(\varsigma_i\left(\tau_i\right)\tau_i\right)^2 \right\rangle = \left\langle \sum_{i=1}^{N}\tau_i^{2\rho_i} \right\rangle$$

(5.7)

where the averaging is over the different samples of the trajectory. The property of the above relations is that the inevitable correlations between increments and the corresponding steps make $\bar{A}$ certainly non-zero. In turn, the latter immediately makes a deviation from the expectation value non-zero. Moreover, Equation (5.6) yields that $\bar{A}$ can become arbitrarily large on increasing N. On the contrary, the independence from one another of the increments and the steps yields $\bar{A} = 0$ which guarantees the uniform convergence of the average to the expectation value. Hence, on a coarse-grained scale and when Equation (5.4) holds, $\bar{A} = 0$ always holds. In turn, it confirms once again that the coarse-grained state space trajectories are BIS. Moreover, the uniform convergence of the average to the expectation value makes the excursion sequence a homogeneous process. We will utilize this property in the next subsection.

Distribution of the Excursions

Invariant Measure

A major consequence of the considerations in the previous sub-sections is that the excursion sequence is a homogeneous process. Indeed, the boundedness and the finite-size memory render uniform convergence of the average to the expectation value of every BIS as it has been established in the previous subsection. In turn, it provides the homogeneity of the excursion sequence. Then, the frequency of occurrence of an excursion of size A is time-independent and reads:

$$P\left(A\right) = cA^{1/\beta(A)}\frac{\exp\left(-A^2/\sigma^2\right)}{\sigma} \qquad (5.8)$$

The required probability $P\left(A\right)$ is given by the duration $\Delta = A^{1/\beta(A)}$ of an excursion of amplitude A weighted by the probability for appearance of excursion of that size (normal distribution). σ is the variance of the BIS; $c = \dfrac{1}{\sigma^{1/\beta(\sigma)}}$ is the normalizing term. The homogeneity of the excursion appearance ensures that $P\left(A\right)$ has the same value at every point of the sequence.

The stationarity of the excursion occurrence proves the assumption that the presentation of a state space trajectory as a sequence of excursions sets the structure of the state space as a dense transitive set of orbits. The transitivity of that set is proven by the following: starting anywhere in the attractor every sequence of excursions reaches every other point in it.

The behavior of any state space trajectory is inherently related to the transition rate statistics through the explicit dependence of $P\left(A\right)$ on $\beta\left(A\right)$. However, as it comes out from (5.3)-(5.4), the increase of the amplitude of excursions, turns $\beta\left(A\right)$ closer and closer to 0.5. Then, $P\left(A\right)$ gradually gets insensitive to the details of the transition rate statistics. Thus, whenever $A >> a$, the behavior of the excursions becomes totally insensitive to it. Then the symmetric random walk appears as global attractor for the fractal Brownian walk regardless to whether it is super- or sub-diffusional. Note, however, that this is true only for bounded fractal Brownian walks, not for arbitrary one!

Thus, so far we have established that the state space is a dense transitive set of orbits. In addition, the trajectories in the state space are BIS. Moreover, on the coarse-grained level these properties are universal and independent on the particularities of the transition rate statistics. Moreover, it has an invariant measure that is given by $P\left(A\right)$ from Equation (5.8). Now we will establish more: we will derive the condition for the asymptotic stability of the invariant measure.

LYAPUNOV EXPONENT OF THE "U-TURNS"

It is obvious that the issue about the asymptotic stability of the invariant measure $P\left(A\right)$ from Equation (5.8) is straightforwardly related to the matter of the "U-turns" defined in Chapter 1. Obviously, the invariant measure is asymptotically stable if and only if the system makes "U-turn" at the approach to the thresholds of stability so that not to involve any specific physical process and/or additional matter/energy. Our task is to derive functional relation among the parameters of a BIS that provides its asymptotic stability.

The requirement that the execution of U-turns should not involve any specific physical process behind it suggests looking for a parallel with the well known phenomenon of the low-dimensional deterministic chaos - phenomenon that occurs at the dynamics of simple deterministic systems. It is associated with unpredictability and great sensitivity to the initial conditions introduced by the mechanism stretching and folding. However, the deterministic chaos also exhibits boundedness: the folding is provided by the fact that the dynamics of the discussed systems is confined to a finite volume of the state space. Along with it, the stretching happens along the unstable directories and gives rise to the unpredictability. Thus, our task becomes to prove that the boundedness automatically introduces a mechanism of stretching and folding.

Our attention is particularly focused on the folding because it sustains the evolution of a chaotic system to be permanently confined in a finite attractor. Intuitively it looks like that the folding in a bounded state space is automatically ensured by the thresholds of stability. However, one may argue that the particularities of the boundary conditions make the folding sensitive to them and hence not universal. The question now becomes to present crucial arguments that the folding is indeed insensitive to the details of the boundary conditions. We consider this problem along with the issue about the folding viewed as a necessary condition for keeping the evolution permanently confined to a bounded attractor. From this point of view, the folding is to be associated with the largest excursions, namely those whose amplitude is of the order of the thresholds of stability.

Below we will find out that the target folding does exist whenever certain relation among the thresholds of stability, a parameter set on the short-range statistics and the variance of a BIS holds. The derivation of that relation involves characteristics of the excursions established in the previous section.

From the viewpoint of the deterministic chaos the folding is to be associated with a negative value of the Lyapunov exponent. Being measure of unpredictability, the latter is the average measure how fast a trajectory deviates under infinitesimally small perturbation of the initial conditions. On the other hand, from the point of view of a BIS, it is to be associated with the largest excursions, namely those whose amplitude is of the order of the thresholds of stability. Thus, our first task is to define the Lyapunov exponent in terms of the excursions and to show explicitly the dependence of its value and sign on their characteristics.

Evidently, the universality of the folding requires that it does not involve any special physical process behind its execution, "U-turns" included. In other words, its implementation must neither involve nor introduce any long-range correlations among the time scales. It has been already established that every coarse-grained BIS has 3 specific parameters: the threshold of stability A_{tr}, the variance σ and the power $\beta(\Delta)$ in the relation amplitude $\leftrightarrow$ duration of the excursions. Next we will prove that when certain relation among these parameters holds, the target folding exists. Note once again that the power $\beta(\Delta)$ in the relation amplitude $\leftrightarrow$ duration is set by the short-range statistics. This justifies our expectation that the target folding neither introduces physical correlations among time scales nor requires additional physical process for the execution of the "U-turns."

From the viewpoint of the deterministic chaos the folding is associated with a negative value of the Lyapunov exponent whose rigorous definition reads:

$$\xi = \lim_{t \to \infty} \frac{1}{t} \ln \left| U(t) \right| \tag{5.9a}$$

where

$$U(t) = X(t) - X^*(t) \qquad (5.9b)$$

$X^*(t)$ is an unperturbed trajectory and $U(t)$ is the average deviation from it. So $|U(t)|$ is the measure of all available deviations from a given point X^*.

On the other hand, from the point of view of BIS, the Lyapunov exponent is to be associated with the excursions since they give rise to essential deviations from the expectation value.

Now we are ready to write down explicitly the asymptotic expression for the average deviation from a trajectory that starts at X^*. The corresponding $|U(t)|$ set on the terms of the excursions reads:

$$|U(t)| = \int_{A^*}^{A_{tr}} AP(A)dA + \int_{A_{cgr}}^{A^*} AP(A)dA \qquad (5.9c)$$

A_{cgr} is the level of coarse-graining, i.e. averaging over all scales smaller than A_{cgr}. This "smoothes out" all the excursions whose size is smaller than A_{cgr} and renders their contribution to the Laypunov exponent zero. Thus, by scanning the value of A_{cgr} we can study only the contribution of the excursions whose amplitude exceed A_{cgr}.

The separation into two terms each of which represents the deviations from A^* to larger and smaller amplitudes is formal. It is made only to elucidate the idea that starting at any point of the attractor one can reach every other through a sequence of excursions. Hence the Lyapunov exponent ξ reads:

$$\xi = \ln \int_{A_{cgr}}^{A_{tr}} AP(A)dA \qquad (5.10)$$

Equation (5.10) proves that value of the Lyapunov exponent for the bounded state space does not depend on an initial point in the state space. In turn, it justifies our assumption that each point in the state space could serve as origin of an orbit.

It should be stressed that the properties established in the previous sections are derived under the condition that all scales larger than the blob size a contribute uniformly to the properties of a BIS. Then the U-turns also do not involve any specific scale larger than the blob size. Thus, the folding being a necessary condition for keeping the evolution of a BIS permanently confined in a finite attractor ensures that the sensitivity to initial conditions produced by stretching and folding is a scale-free process.

It is to be expected that the size of an excursion determines its contribution to the stretching or folding. Indeed, since the frequency of the small size excursions is essentially high, figuratively speaking, they "hold" every trajectory permanently deviated from the expectation value. So, the small size excursions most probably contribute to the stretching of the trajectories. On the contrary, the largest excursions are rather occasional and the corresponding trajectory spends most of its time as close as possible to the expectation value. So, they contribute rather to the folding. The explicit revealing of the role of small and large excursions is made by the use of coarse-graining: the role of the excursion size is carried out by scanning the ratio A_{cgr}/σ. The ratio A_{cgr}/σ has two extreme cases:

1. $\dfrac{A_{cgr}}{\sigma} \ll 1$, i.e. the contribution of the small excursions prevails. By the use of the steepest descent method, Equation (5.10) yields:

2. $\xi \approx \ln \sigma$ (5.11)

3. Equation (5.11) tells that asymptotically every trajectory visits every point in the attractor so that the mean deviation from the initial point is the same for every trajectory

and is bounded by the thresholds of the attractor itself. The positive value of ξ justifies our speculation that the small size excursions contribute predominantly to the stretching. Further, visiting of every point of the attractor starting anywhere in it makes the motion on the attractor ergodic.

4. $A_{cgr} >> \sigma$, i.e. large scale excursions contribution prevails. Equation (5.10) yields:

5. $$\xi \approx \left[\frac{1}{\beta\left(A_{cgr}\right)} + 1 \right] \ln \frac{A_{cgr}}{\sigma} + \ln \sigma - \frac{A_{cgr}^2}{\sigma^2}$$
    (5.12)

While ξ from Equation (5.11) is always positive which provides stretching, Equation (5.12) opens the alternative for ξ being both positive or negative depending on the relation among $\beta\left(A_{cgr}\right)$, σ and A_{cgr}. Then, since its natural measure is the negative value of the Lyapunov exponent, the folding that permanently keeps the evolution bounded in a finite attractor is provided if and only if A_{tr}, $\beta\left(A_{tr}\right)$ and σ are such that $\xi < 0$:

$$\xi \approx \left[\frac{1}{\beta\left(A_{tr}\right)} + 1 \right] \ln \frac{A_{tr}}{\sigma} + \ln \sigma - \frac{A_{tr}^2}{\sigma^2} < 0$$
(5.13)

We should immediately admit that Equation (5.13) makes the realization of the "U-turns" automatic. This renders the folding insensitive to the details of the thresholds of stability and the way they are approached. That is why we call the condition (5.13) Lyapunov exponent of the "U-turns."

It should be stressed once again that the power $\beta\left(A\right)$ in (5.13) is set by the short-range statistics. Then, the "U-turns" neither require nor introduce long-range physical correlations among the time scales. This result agrees with the fundamental assumption about the uniform contribu-

tion of all time scales. In the present context it implies lack of any physical processes, "U-turns" included, that yields long-range physical correlations among the time scales.

It is worth noting that the folding is a broader notion than the tangent approach to the boundary. Both folding and the tangent approach produce the same effect: they contribute to the convergence of a trajectory making it to depart from the threshold. Yet, the tangent approach itself is a property of the random walk that creates the excursions entailed with the appropriate boundary conditions, while the folding is provided by Equation (5.13) without any implication of the particularities of the boundaries.

POWER LAWS UNDER BOUNDEDNESS

In section *Distribution of Excursions. Invariant measure* we called the probability distribution $P\left(A\right)$ for the excursion occurrence an invariant measure. However, the rigorous definition of an invariant measure asserts that a distribution appears as an invariant measure only when it retains the property to be preserved by some function. However, it is obvious that there is not such function that exactly preserves $P\left(A\right)$ from (5.8). Then why we insist on calling it invariant measure?! It is because next we will prove that $P\left(A\right)$ from (5.8) stands as invariant measure in an approximation; an exclusive property of that approximation will be that it excellently matches a power law distribution. A distinctive property of that approximation is that it successfully releases the inconsistency between the scale invariance of the power laws and the time translational invariance of the distribution $P\left(A\right)$ from (5.8).

We start our considerations by presenting brief information about the power laws distributions. A power law is a special kind of mathematical relationship between two quantities. When the frequency of an event varies as a power of some attribute of that event (e.g. its size), the frequency is said to follow a power law. For instance, the number of cities having a certain population size is found to vary as a power of the size of the population, and hence follows a power law. The distribution of a wide variety of natural and man-made phenomena follow a power law, including frequencies of words in most languages, frequencies of family names, sizes of craters on the moon and of solar flares, the sizes of power outages, earthquakes, and wars, the popularity of books and music, and many other quantified phenomena. It also underlies the "80/20 rule" or Pareto distribution governing the distribution of income or wealth within a population. A power law is any polynomial relationship that exhibits the property of scale invariance. The most common power laws relate two variables and have the form

$$f(x) = ax^k + o(x^k) \tag{5.14}$$

where a and k are constants, and $o(x^k)$ is an asymptotically small function of x^k. Here, k is typically called the *scaling exponent*, where the word "scaling" denotes the fact that a power-law function satisfies $f(cx) \propto f(x)$ where c is a constant. Thus, a rescaling of the function's argument changes the constant of proportionality but preserves the shape of the function itself.

Thus, the scaling invariance of Equation (5.14) provides the power laws to appear as invariant measures. Yet, there are 3 still puzzling facts about the power laws:

1. The ubiquity of power laws distributions: A few notable examples of power laws are the Gutenberg-Richter law for earthquake sizes, Pareto's law of income distribution, structural self-similarity of fractals, and scaling laws in biological systems. In the Preface of the book we already discussed another

notable example of power law, the Zipf's law. Then, how the variety and diversity of these phenomena is incorporated in a single type of distribution?!

2. The inconsistency with the time translational invariance: the central idea behind the traditional approach to the power law distributions is that their onset is a diffusion process the memory size of whose walk is unspecified. The most pronounced manifestation of that property is the scale invariance of the power law distributions. Alongside, the same distributions are assumed time translational invariant, i.e. their scale invariance is supposed independent from the beginning of the diffusion process. Yet, the concurrence of the both properties is biased by the conflict between the circumstances for their performance: while the scale invariance implies spanning of a power law over finite time intervals of unrestricted size, the time translation invariance requires the power law to be defined on infinitesimal time intervals only;

3. The third weak point concerns the conditions for sustaining long-term stable evolution. Obviously, a self- sustainable evolution is established only when the rate of exchanging energy/matter with the environment is kept permanently fixed. However, a power law distribution implies that the rate of exchanged by the stochastic deviations energy/matter does not converge to a steady value but gradually increases with the length of time window. Then, it would turn out that the systems subject to power law distributions are unstable. The expected instability, however, is in a strange contradiction with the apparent empirical stability of the systems where it has been observed.

We start the revelation of the above puzzle by shedding light on the problem about the concurrence of the time-translational and the scale invariance of the distribution (5.8). Our expectations are that scale invariance is implemented by the highly non-trivial impact of the correlations induced by "U-turns" on the memory size: the inevitable repetition of "U-turns" in a finite time interval efficiently enlarges the memory size, initially constrained by the dynamics to the size of a blob. On the other hand, the same "U-turns" keeps the size of the excursions bounded and thus acts towards sustaining the leading role of the time-translational invariance. Since the scaling invariance of the boundedness neither introduces nor selects any specific time scale, we assume that power law distributions are a good approximation to the genuine distribution (5.8). Next we will illustrate that the probability (5.8) has a heavy tail that is excellently fitted by a power law. Further, we will demonstrate that this fit helps to elucidate another famous property of a complex behavioral pattern: "stationarity of increments." Yet, the "thin ice" is the issue about the margins of feasibility of power function approximation.

The time-translational invariance of the probability for an excursion, given by (5.8), becomes evident when the size of the excursion is expressed through the following relation:

$$P\left(\Delta\right) \propto \Delta^{\beta\left(\Delta\right)} \exp\left(-\frac{\Delta^{2\beta\left(\Delta\right)}}{\sigma^2}\right) \qquad (5.15)$$

where the duration Δ appears as an independent variable, namely it represents the length of the time window. Then, (5.15) indicates that the probability for having an excursion depends only on the length of the window but is independent on the location of that window on time arrow: just what time-translational invariance implies.

A plot of $P\left(\Delta\right)$ with $\beta = 0.3$ and $\sigma = 2$ is presented in Figure 2 (black line). The heavy tail is fitted by the power function $f\left(\Delta\right) \propto \Delta^{\gamma\left(\Delta\right)}$ with $\gamma\left(\Delta\right) = -0.1 - 0.005\Delta$ (dotted line). The most impressive result of this approximation is that the

surprisingly good fit is spanned over two orders of magnitude and covers the entire tail of the distribution $P\left(\Delta\right)$.

The success of the heavy tail approximation with a power law function justifies all empirically obtained relations and characteristics such as stationarity of the increments, introduction of Hurst exponent, power dependence of the moments etc. Thus, the stationarity of the increments is an immediate consequence of the relation between the size and duration of an excursion (5.15) (Box 1):where the last two lines appear in result of approximating the heavy tail of

$$\Delta^{\beta\left(\Delta\right)} \exp\left(-\frac{\Delta^{2\beta\left(\Delta\right)}}{\sigma^2}\right) \text{ by } f\left(\Delta\right).$$

Further, the introduction of generalized Hurst exponent as "index of dependence" that measures the relative tendency of a time series to either strongly regress to the mean or "cluster" in a direction is justified by the coarse-grained structure of a BIS that is a succession of excursions each of which is embedded in an larger time-interval. Then, the excursions themselves are the intervals of "clustering" while the embedding time intervals that separates each two successive excursions are the intervals of regress to the mean. Thereby, the value of Hurst exponent is to be associated with $\beta\left(\Delta\right)$ from (5.15).

Yet, in the present setting all characteristics and relations introduced by the developed so far approaches to the matter of power laws have an entirely novel understanding and some of them appear as only approximations with limited feasibility. For that reason it is important to outline the margins of that feasibility. The first non-trivial example is the issue about the moments. The boundedness concept suggests that a bounded time series has only two steady characteristics: expectation value and variance. Then, its higher moments are to be dependent only on them. Indeed,

Figure 2. Approximation of the probability for an excursion (continuous line) with power function (dotted line)

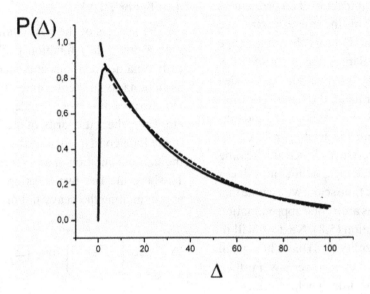

taking into account (5.15), the $n-th$ moment reads:

$$\langle A^n \rangle \propto \int_0^{A_{max}} A^n P(A) dA = \int_0^{A_{max}} A^{n+1/\beta(\Delta)} \exp\left(-\frac{A^2}{\sigma^2}\right) = \sigma^{n+1/\beta(\Delta)}$$

(5.17)

Indeed, it depends only on the variance but is independent from the length of the time-window on the contrary to the traditional approach involving a power law distribution. Yet, if suppose that, because of its dominant role, only the heavy tail contributes to the distribution:

$$\langle A^n \rangle \propto \int_\sigma^{A_\Delta} A^n f(A) = \int_0^{A_\Delta} A^{n+\gamma} dA \propto A_\Delta^{n+\gamma+1} \propto \left[\left(\Delta\right)^{\beta(\Delta)}\right]^{n+\gamma+1}$$

(5.18)

where Δ is the length of the time window. Then, the $n-th$ moment appears proportional to the length of the window on the contrary to Equation (5.17)?! This paradox is easily resolved by the fact that Equation (5.18) holds only for restricted window sizes, namely for windows whose sizes are smaller than the duration of the largest excursions (those whose size equals the thresholds of stability). The reason for this constraint is that on

Box 1.

$$\langle A(t+\Delta) - A(t) \rangle \propto \int_{-t_{max}}^{t_{max}} (t+\Delta)^{\beta(t+\Delta)} \exp\left(-\frac{(t+\Delta)^{2\beta(t+\Delta)}}{\sigma^2}\right) - \int_{-t_{max}}^{t_{max}} t^{\beta(t)} \exp\left(-\frac{t^{2\beta(t)}}{\sigma^2}\right) \propto$$

$$\propto \Delta^{\beta(\Delta)} \exp\left(-\frac{\Delta^{2\beta(\Delta)}}{\sigma^2}\right) \propto f(\Delta) = \Delta^{\gamma(\Delta)}$$

(5.16)

reaching the thresholds of stability it is not longer feasible to ignore the "U-turns" whose role in the present context is to "switch" the distribution from the tail to its beginning and thus makes the approximation of the genuine distribution $P(A)$ with the power function $f(A)$ inappropriate. Outlining, the power function fit is good approximation which cannot be spread to window of arbitrary length because the approximation does not take into account the "U-turns." Thus, the approximation of the genuine distribution (5.15) with a power function is limited by the duration of the largest excursion. Yet, as seen from Figure 2 the heavy tail is spanned on the two orders of magnitude; taking into account that this span involves already made coarse-graining, the actual span could be of several more orders of magnitude. Thus, the heavy tail of the distribution of excursions $P(A)$ indeed excellently fits the scale invariance of the empirical power laws.

Further in this line of reasoning comes the fact that the genuine distribution of excursions $P(A)$ from (5.8) is normalizible so that the normalizing coefficient does not depend on the size of the window. Therefore, even though the exchanged matter/energy in a fluctuation varies from one excursion to the next, the exchanged amount of energy/matter uniformly converges to a steady value which is independent from the size of the time window. In turn, the uniform convergence provides permanent stability of the system.

Thus, we prove the dominant role of the time-translational invariance of the hierarchical self-organizations as a major characteristic of the stability of a system. Alongside, the scale invariance appears as a surprisingly good approximation to the distribution of excursions. Keeping in mind that the distribution of excursions $P(A)$ from (5.8) is equally available for intelligent and non-

intelligent systems, one can easily explain the ubiquity of the power law distributions. It is worth noting that it is non-trivial expression of the non-physical correlations among distant responses introduced by the static and dynamical boundedness and thus it stands as an exclusive property of the boundedness as a conjecture whose primary task is providing the time translational invariance of a system. In turn, the ubiquity and universality of the power law distributions viewed as an exclusive excellent fit to the invariant measure (5.8) render their utilization as a universal criterion for hierarchical structuring under boundedness.

CONCLUSION

One of the major goals of the concept of boundedness is to explain the properties of the complex systems in a self-consistent way. The greatest challenge to modern interdisciplinary science is the ubiquitous coexistence of specific and universal properties. The enigma of this coexistence consists in a confrontation between the empirical stability of the systems, and its violation commencing from its description. Notable examples of universal characteristics that drive the above contradiction are: power spectra and power law distributions. For a long time both are supposed to fit power dependent shapes. However, any power dependence yields either infrared or ultra-violet catastrophe on its integration. Therefore the averaged energy/matter exchanged with the environment and the size of the fluctuations gradually enlarges with the time thus driving to the collapse of a system. Thus we face the dilemma whether the shape is exactly a power dependent one or there is something different. What is more important is whether the deviations from a strict power dependence could be put in a self-consistent general frame which reveals the ubiquitous coexistence of specific and universal properties. Alongside, we should answer the question why both intelligent and non-intelligent systems share the same behavior.

We have put forward the concept of boundedness in an attempt to set it as desired frame for explaining the above presented puzzling problems. Its major focus is put on the stability of complex systems so that the latter to be the central property from which all others are derived. Thus, the concept of boundedness asserts that the energy/matter accumulation and the rate of its exchange at any level of matter organization are bounded to be within specific margins; the response is local and not pre-determined. Thus, the response is free from being a subject of any apriori given distribution. Then, we have proved that under the minimal condition of retaining metrics, the power spectrum comprises continuous band whose shape is $1/f^{\alpha(f)}$ and which is independent from the statistics of the fluctuations. The major value of this proof is that the obtained shape provides convergence of the integral over it in any time window. Thus, we prove that exchanged matter/energy is constant and independent both from the size of the window and its position on the time scale. Thus, we proved that the persistence of a continuous band in the power spectrum could be in accordance with the stability of the system when reading of the fit is slightly modified.

The above considerations suggest a crucial test for the concept of boundedness: whether it could produce power laws distributions. On the one hand, it seems that the idea of universal distribution derived from the concept of boundedness is rather an absurd because the concept of boundedness assumes the response to be local and not pre-determined and thus the fluctuations are free from being subject of any global distribution. On the other hand, the boundedness introduces non-physical correlations among distant responses. Our expectations are that scale invariance is implemented by the highly non-trivial impact

of the correlations induced by "U-turns" on the memory size: the inevitable repetition of "U-turns" in a finite time interval efficiently enlarges the memory size, initially constrained by the dynamics to the size of a blob. On the other hand, the same "U-turns" keeps the size of the excursions bounded and thus acts towards sustaining the leading role of the time-translational invariance. Since the scaling invariance of the boundedness neither introduces nor selects any specific time scale, we assume that power law distributions are an excellent approximation to the genuine distribution (5.8). It is worth noting that the genuine distribution of the excursions (5.8) is derived under the assumption about the leading role of the time-translational invariance. And indeed, the plot in Equation 5.2 demonstrates that the distribution (5.8) has a heavy tail which is excellently fitted by a power law function spanned over several orders of magnitude.

Beyond doubt we have achieved our goal to derive self-consistently the major properties of the complex systems whose description seems to challenge their empirically established stability. Thus, we have managed to reconcile the time-translational invariance with the scaling invariance so that the latter to acquire the following understanding: the scaling invariance appears as a property that no specific time/space scale has particular contribution to the universal properties of the complex systems at any level of matter/energy organization. It is worth noting the decisive role of the exclusive for the boundedness property of multi-valuedness of the relation between the size of an excursion and its embedding time interval; since a set of embedding time intervals of different duration is assigned to each excursion, the obtained multi-valuedness renders a random choice of one embedding time interval to each excursion at each of its occurrences. In turn this breaks any accidental long-term correlations among distant excursions which in turn ensures ban over appearance of extra lines in the discrete band of the power spectrum. Thus, the "homeostasis" is maintained robust in an ever changing environment.

Yet, our intuition remains unsatisfied: why intelligent systems also share the property of obeying power law distributions?! The problem is particularly serious for Zipf's law which asserts that: given some corpus of natural languages, the frequency of any word is inversely proportional to its rank in the frequency table. Thus the most frequent word will occur approximately twice often as the second most frequent word, three times as often as the third most frequent word, etc. Put it in other words, the Zipf law ignores any semantic meaning and thus seems to sweep out the difference between mind activity and random sequences of letters. It is a common knowledge that the semantics is permutation sensitive and puts a long-range order among its units. Then, how is it possible to occur that these long-range correlations can be ignored? The answer is not easy and comes only after a closer look at the structure of a semantic sequence. We start with reminding that under the concept of boundedness, any semantic sequence is a sequence of orbits in the corresponding state space. The permutation sensitivity is implemented by associating the meaning of each orbit with the functional irreversibility of a specific engine. Thereby, running the orbits each in its direction substantiates specific long-range correlations which are set by the semantics. However, when considering the frequency of occurrence of a given word in a text, we ignore these higher level correlations and reduce the "system" to its simplest counterpart, i.e. to the first hierarchical level; and as we already have established, its state space is a dense transitive set of orbits subject to power law distribution.

Thus, we obtain a remarkable result: our "intervention," i.e. the way we study a system strongly influences the obtained result! Does this imply that the "measurements" of hierarchically structured systems ignore certain correlations?! We are not ready yet to answer this question rigorously, but still we are able to pose it. The case of the Zipf's

law strongly suggests that any "measurement" of hierarchical order inevitably breaks its functioning: thus, the "measuring" of the frequency of occurrence of a word in a text ignores its position in the text and thus ignores the "order" set by the semantics. Therefore, we come to the conclusion that any measurement "reduces" a system to a lower level counterpart. This poses a number of questions: the first one is whether it is ever possible to make measurement that does not "reduce" the hierarchical order; if not, is there a general objective criterion for defining how deep is the process of reduction in any particular measurement? We still are not able to give a general answer to the above questions. Yet, we could firmly conclude that the association of the semantic meaning with the functional irreversibility of specific engines makes any "measurement" in terms of frequencies for occurrence inevitably belonging to this class. In turn this renders the traditional probabilistic approach "reductionist" in the sense that its involvement is always supplemented by certain "reduction' of the studied system to a lower level counterpart. It should be stressed that taking into account the "reductionism" of the traditional probabilistic approach is crucial for the study of the intelligent systems.

It should be stressed that the above considerations clearly demonstrate how crucially the comprehension of measurements we made depends on the current views on the matter. Indeed, we encounter two fundamentally different settings for power law distributions: (i) the traditional approach assumes the accuracy of power law distributions at the expense of letting their confrontation with time-translational invariance and empirically observed stability; (ii) our approach considers the power law distributions as a result of "destructive" measurements which, however, retains the exclusive property to serve as a universal criterion for hierarchical structuring under boundedness. In turn, the well developed techniques of the probabilistic approach acquire new role: to help establishing hierarchical structuring.

Chapter 6
Boundedness and Other Theories for Complex Systems

ABSTRACT

A comparison between the concept of boundedness on the one hand, and the theory of self-organized criticality (SOC) and the deterministic chaos on the other hand, is made. The focus is put on the methodological importance of the general frame through which an enormous class of empirical observations is viewed. The major difference between the concept of boundedness and the theory of self organized criticality is that under boundedness, the response comprises both specific and universal part, and thus a system has well defined "identity," while SOC assumes response as a global invariant which has only universal properties. Unlike the deterministic chaos, the boundedness is free to explain the sensitivity to initial conditions independently from the mathematical object that generates them. Alongside, it turns out that the traditional approach to the deterministic chaos has its ample understanding under the concept of boundedness.

INTRODUCTION

We have developed the concept of boundedness for explaining in a self-consistent way the properties of complex systems. The goal of the present chapter is to compare the concept of boundedness with two established approaches to the subject-matter: self-organized criticality and deterministic chaos. The purpose is not to make a critical review but rather to demonstrate the fundamental differences between each of them and the concept of boundedness. By means of substantiating this problematic in such a way we hope to demonstrate not only the merits of our approach, but to focus

DOI: 10.4018/978-1-4666-2202-9.ch006

on the methodological importance of the general frame through which we view an enormous class of empirical observations. Such a general frame offers unique choice that combines general view and logic and turns out crucial for substantiating an approach with far going consequences. The high non-triviality of the matter is induced by the range of its application: it encompasses an enormous diversity of phenomena and systems each of which belongs to a different branch of science. It is believed that in the different scientific domains different rules and laws operate so that any systematic approach to them starts with the question: is there a general interdisciplinary rule which generates that peculiar behavior? If the answer is positive then it implies a coexistence of specific and universal properties.

Our answer to this question is affirmative: we assert that there exists a general approach based on the concept of boundedness. The theoretical background of this concept does not allow it to be considered as a law in its traditional understanding. It assumes an operational protocol. The fundamental difference between the idea of operational protocol and the traditional understanding of a law is that the former neither specifies the environment nor requires its exact reoccurrence along with the reoccurrence of a given phenomenon. To remind, the traditional understanding of a law is set on establishing certain, usually quantitative, relations among a number of variables so that the relation to be invariant on reoccurrence of the event. These considerations entail the strong pre-supposition that the local environment of the event reoccurs also with the reoccurrence of the event. Instead, we assume that there exists certain quantitative relations which stay invariant even in an ever-changing environment. The difference with the traditional approach is that our assertion drives us closer to the notion of homeostasis than to the idea of thermodynamical equilibrium. It should be highlighted that we understand the notion of homeostasis in its most general meaning: it is the property of a system, either open or closed, to regulate its internal environment and tends to maintain a stable, constant condition. Thereby, we have made two fundamental steps: (i) the first one consists of imposing the notion of homeostasis, borrowed from biology, in explanation of the behavior of physical systems which are supposed to be sufficiently described by the idea of thermodynamical equilibrium. To remind that thermodynamical equilibrium is the state associated with the maximum entropy and which is a global attractor for every initial condition; on the contrary, the homeostasis is self-organized pattern which is robust to environmental fluctuations in a certain domain of control parameters; thus it is neither the state of maximum entropy nor its properties are explicitly related to the notion of entropy; moreover it is released from the requirement to be a global attractor and thus opens the door to "adaptation" of homeostasis on changing domains of control parameters. (ii) the second step consists of adopting the idea that the matter organization has hierarchical structure which goes in both directions: bottom up and top down so that there exists general operational protocol that governs it.

The idea that the complex systems are out-of-equilibrium is not new. Actually it is one of the more advanced and elegant ideas put forward for explaining the behavior of complex systems. Yet, there is a fundamental difference with the idea of homeostasis. Essentially, it involves the idea of out-of-equilibrium that goes in accordance with the general view for "bottom up" matter organization. Thus, the complex systems are considered as a class of systems whose behavior is governed by the laws established at the lower levels of matter organization. The great breakthrough has come by the theoretical and experimental discovery of the emergent phenomena based on the pioneering work of Belousov, Zhabotinsky, Turring and Prigogine, to mention a few. They established that, tuned to a certain range of control parameters, some very simple systems tend to self-organize spontaneously into spatio-temporal patterns.

The major property of these patterns is that their creation substantiates a strong contradiction with the Second Law, a law which is proclaimed as one of the basic laws of Nature. Thus, we face a dilemma: whether there are general rules available to each level of matter organization and whether the Second Law can be reformulated so that to be one of these rules? The role of the Second Law is vitally important for open systems since they permanently exchange matter/energy with the environment. Then, it is important to establish whether the exchanged matter/energy is constant on average or it gradually changes. In order to give answer to this question we need to ask another one: whether a system is in balance with the environment – this is the case when the exchanged energy/matter is constant on average; if so it is to be expected that the interaction between a system and its environment will not allow permanent "drainage" of the environment for creating "work". Put in other words, it will not allow a perpetuum mobile. Then, if proven as a general result, the ban over a perpetuum mobile will serve as the desired reformulation of the Second Law. Thus the ban over perpetuum mobile will come to replace the most widely used so far formulation of the Second Law. According to the traditional formulation of the Second Law, the equilibrium is the state of maximum entropy which is global attractor for all initial states.

The greatest difficulty and non-triviality of the subject-matter lies in the traditional reading of the empirical observations that appears in the shape of the continuous band in the power spectra and the power laws distributions. On the one hand, the ubiquitous robustness of the obtained power dependences strongly suggest the idea that all the diversity of systems shares the same organizational principle; on the other hand, it puzzles by the fact that any power dependence implies instability of a system due to gradual changes in the matter/energy exchange with the environment. Thus, the gradual changes of the exchanged matter/energy not only allow "drainage" of the environment

but imply total instability of the Universe and thus imply permanent mix of Big Bang and Big Crunch. The implausibility of this line of reasoning is suggested by the important and ubiquitous confrontation of two general empirical facts: the ubiquity of the power dependencies on the one hand, and the remarkable stability of the corresponding systems on the other hand. Thereby we face another fundamental problem: how accurate are the empirical observations and whether we read them properly; do different readings bring about different results? Our further consideration will reveal that indeed, the obtained empirical data are extremely sensitive to their reading and their embedding in the corresponding explanatory concept. This will be demonstrated trough the merits and the drawbacks of the corresponding approaches.

BACKGROUND

The relation determinism-stochasticity appears to be critical for adoption of any approach to complex systems. The traditional understanding of these notions is as follows: the determinism is associated with single-valued relations which can be set and measured with arbitrary precision. Thus the hallmark of the determinism is the relation single cause – single effect. On the contrary, the stochasticity is associated with lack of any cause-effect relation: it is associated with random choice of a single outcome from a set of many ones. This view has its strong support set by the Central Limit Theorem which states that given a set of random independent events which has finite variance, they retain asymptotic distribution which is the normal one. The power of this result is that it seems to prove the central for the statistical physics result that the larger fluctuations are less probable. Further, the assumption about the independence of the fluctuations from one another is achieved through the supposition that each of them starts and ends at equilibrium which is a global attractor.

However, the properties of the complex systems cast serious doubts to the above scheme. First of all, the coexistence of specific and universal properties draws the traditional understanding of the relation determinism-stochasticity into serious contradiction. According to this contradictory bias it seems natural to assign a deterministic origin to the specific properties and a stochastic one to the universal properties. But a question arises: does this natural assumption imply that the behavior of a complex system always splits into two independent parts?! In order to make this problematic situation more clear we have to reformulate the question to the following one: how a system which consists of two independent parts is put together and operates steadily as a single object; how its ingredients "know" to which part they belong and how to behave accordingly; how the separation is implemented at molecular level?

Further in this line of reasoning comes the question about the scale invariance of the power laws. Let us assume that there is an exact observation. It implies that on averaging the distribution remains the same. This property rather suggests presence of long-range correlations among the distant fluctuations than their independence. To remind, the normal distribution is not scale invariant because it is supposed to be constituted by independent events! Moreover, the scale invariance implies certain relation between the position of a fluctuation and its neighbors. Is it a stationary process and if so do the correlations suggest certain determinism in the succession of fluctuations or it calls for a fundamentally new understanding of the relation determinism-stochasticity? Do the correlations commence from a specific physical process and if not where do they come from? And since they are still described by a probability distribution what is their radius and if it is finite what are the consequences this property signals for?

An immediate result from the above cascade of puzzling questions is that the response of the complex systems is active: the correlations among neighboring responses are sign that the corresponding system "reacts" actively to any external impact so that to self-organize its internal environment for maintaining its state intact. Thus, we come back to the hypothesis of replacing the thermodynamical equilibrium with the idea of homeostasis. This step is crucial since it opens the door to diversity of specific property by means of allowing the homeostasis to be associated with every appropriate spatio-temporal pattern, not with the single state of maximum entropy. It should be stressed that the thermodynamical equilibrium cannot provide the empirically observed huge diversity of specific properties.

The assumption about the homeostasis poses the question why the response is not entirely deterministic, but has a component which is, to a certain extent, proportional to the impact. Why a system needs the separation of its response to a robust (active) and a passive component? This question comes in pair with the question whether the power law distributions imply higher probability for rare events and thus confronts our intuition that a condition for the stability is the fulfillment of the relation: the larger fluctuations are less probable. It is worth noting that these considerations are highly non-trivial since the correlations among successive fluctuations established through power law distributions violates the condition for independence of the events participating in constituting the Central Limit Theorem. Thus, our quest for independent events in the form of excursion turns remarkably successful: the results of the previous Chapter prove the idea that the power law distributions are just remarkable approximation to the genuine distribution (5.8). It is worth noting that the distribution (5.8) is not scale-invariant! Yet, it confirms our intuitive assumption that the larger events are less probable. Thereby the properties of scale invariance and higher probabilities for rare events appear as properties of the approximation to the heavy tail of the distribution (5.8). Thus the explanation for the appearance of these properties has to be sought in the highly non-trivial role of the non-physical correlations among distant responses imposed by the boundedness.

The above considerations set an example for the extent of diversity of approaches and views on complex systems which have far going consequences. The fundamental importance of the above considerations is that they are an example for the sensitivity of the process of reading the empirical data relative to the context of an attempt for overall conceptualization of the subject-matter. Thus the line: scale invariance of power law distributions $\Rightarrow$ higher probability for rare events $\Rightarrow$ dominance of stochasticity drives nowhere because it confronts with the idea of correlations among distant fluctuations leaving their radius unspecified which in turn violates time-translational invariance On the other hand, the line: stability $\Rightarrow$ boundedness $\Rightarrow$ self-organization and hierarchical order $\Rightarrow$ excursions $\Rightarrow$ power law distributions as an approximation ensures the condition that the larger fluctuations are less probable and thus provides time-translational invariance.

CRISES OF IDENTITY AT THERMODYNAMICAL EQUILIBRIUM

One of the central outcomes of our reading of the properties of complex systems is that their response is active, i.e. a system self-organizes its response so that to retain its current state intact as much as possible and at the same time to "react" to the tiniest environmental impacts. So, the "homeostatic" component of the response retains the specific properties of the system. Thus we face the methodological question to what extent we should use the traditional statistical mechanics. One of the main notions in the traditional statistical mechanics is that of the thermodynamical equilibrium is viewed as the state of maximum entropy. This view comes out from the conjecture of the

Second Law of thermodynamics as an expression of the tendency that over time, differences in temperature, pressure, and chemical potential equilibrate in an isolated physical system. The law deduces the principle of increase of entropy to explain the phenomenon of irreversibility in Nature. Even though the Second Law in the present formulation is widely used, next we will prove that it is inconsistent with the idea of active response.

Turning point in our considerations is an overlooked so far inconsistency of the first and the second laws of thermodynamics with the notion of a closed system. Our task is to demonstrate that this discrepancy has far going consequences: it attacks the very idea that a system in equilibrium can be ever isolated from its environment. We start revealing that conflict by reminding that the bridge between the first and the second law is grounded on the following relation between heat and entropy:

$$dQ = TdS \qquad (6.1)$$

where Q is the heat, T is the temperature and S is the entropy. Together with the first law of the thermodynamics:

$$dU = dQ + dW \qquad (6.2)$$

(6.1) points out that the entropy approaches its maximum at the expense of exerting transformations "work" $\rightarrow$ heat $\rightarrow$ entropy; here U stands for internal energy and W for "work". So far it has been taken for granted that this sequence of transformations is to be interpreted as follows: the action of external macroscopic "forces" is transformed to energy dissipation on dynamical level which in turn brings about spatial rearrangement of the system such that the entropy reaches its maximum at equilibrium. At the same time Q and W are considered independent from one another functions. However, it turns out that both assertions are inconsistent and their conflict generates the following insurmountable difficulties:

- Since Q and W are considered independent, the assumption that the entropy reaches maximum at equilibrium imposes constraint only on Q (requiring $dQ \equiv 0$) whilst dW remains arbitrary. In turn, (6.2) reduces to $dU = dW$ which, however, implies that the internal state of a system, characterized by U, becomes explicit function of the environmental circumstances characterized by W. Therefore, each time a *closed* system reaches the entropy maximum, its identity has become lost because its behavior is immediately controlled by the environment?! Besides, adding the condition that the thermodynamical equilibrium requires not only maximum of the entropy but maximum of the internal energy as well does not help to improve the situation because generically the maxima of U and Q do not coincide.

- The laws (6.1)-(6.2) are incomplete to provide monotonic increase of the entropy on its approach to equilibrium. Let us suppose that a system is subject to constraints that generate some spatial heterogeneity; as an immediate consequence, the entropy becomes spatially dependent. Accordingly, the global entropy maximum, if ever reached, is not anymore a global attractor because some paths go via local minima the escape from which, however, is achieved at the expense of a non-monotonic change of the entropy.

Summarizing, it becomes obvious that the traditional formulation of the Second Law suffers severe flaws and must be seriously reconsidered. That is why our top priority is to find out the relation between the proper identity of a system and the constraints to which it is exposed so that this identity *not* to be immediately controlled by the environment.

The above considerations demonstrate that not only the idea for active response is inconsistent with the traditional notion of thermodynamical equilibrium but it turns out that the latter is inconsistent with its traditional subject – closed systems. However, the rejection of the traditional formulation of the Second Law does not make our task easier: now we should look for an approach that is equally available for both open and closed systems. One of the central tasks of any unifying approach is the explanation why the same system, e.g. a chemical reaction, behaves differently when being closed and open? To explain, when closed, a system always behaves irreversibly, i.e. it tends to reach a single steady state where differences in temperature, pressure, and chemical potential are equilibrated; on the contrary, when being open, the same system behaves so that to its properties are closer to those of a complex system: it can reside at more than one stationary state; the properties of the stationary states are governed by an active interplay between the internal dynamics and the control parameters. Then, why at certain conditions (closed systems) the response is inactive while at others (open systems) the response of the same system turns "active"?

Next in this line of considerations comes the question whether a frame, general enough to include both open and closed systems in a self-consistent way, is possible to exist. This is not an easy task since it inevitably goes through the dilemma whether there is such reformulation of the Second Law able to be equally available for both open and closed systems; and not only for them: our considerations suggest that semantics is executed by means of natural processes. Then, the semantics also must be subjected to the Second Law. Then, among all, we face the question whether there exists a universal measure for the meaning of any semantic unit and if so how to distinguish the one semantic unit from any other. By generalizing the subject-matter this question speaks about the "energetic" price of the information. Note that it is not reduced to a number of bits, size of brain or any other quantitative criterion since, as we have already demonstrated that the semantic is inherently related to the notion of hierarchical order, it is related to a hierarchically organized engines. Thereby, the idea of self-organized semantics has its apparent relation to the idea of the Second Law viewed as a ban over perpetuum mobile. Yet, our suggestion needs a rigorous proof because it still confronts not only the separation of the systems to open and closed but it confronts the empirical observations about the power law distributions established for complex systems. To remind, power dependence implies that the corresponding energy/matter involved in the corresponding system gradually increases with time; alongside, the variance of the fluctuations also gradually increases. Thereby, the system would "drain" its environment by means of developing "rare events", i.e. by means of making larger fluctuations more probable. Thus, the system would experience gradual changes and would tend to collapse in a finite period of time. However, noting like this happens. Thus we face another important dilemma: how reliable are the empirical observations?

The crises of the traditional formulation of the Second Law viewed as a principle available at every level of matter organization has deeper origin: it appears already at quantum level. To explain: as we have demonstrated in Chapter 3, the local instability turns spontaneously amplified when considering all dynamical interactions as unitary ones. Keeping in mind the well known assumption that in statistical thermodynamics, the second law is a *consequence* of unitarity in quantum theory, our result about unrestrained amplification of the local instabilities breaks the fundament of the statistical theory. However, the role of the unitarity at quantum level is highly controversial since it is "responsible" for another non-overwhelming difficulty in the statistical mechanics: that of the role of time-reversal symmetry of any dynamical process. Indeed, the time reversal symmetry renders any process equally available with its reverse:

in turn this makes equally likely that a small drop of ink dissolved in water, spontaneously "gathers" again into a drop from the homogeneous solution. Then, why this never happens?!

The concept of boundedness gives credible answer to the latest question: this never happens because the overlooked non-unitary interactions that we have introduced in Chapter 3 act always in one direction, namely towards decreasing of the local instabilities. They act as permanently propelling "mixers" whose role is to maintain the homogeneity of the system. It is worth noting, that these interactions acquire the property of being non-unitary under the assumption that the potential and the random motion are *not* separable, a property which is basis for the unitarity in the traditional quantum theory. Therefore, it turns out *impossible* that, once dissolved into water, a drop of ink "gathers" again from a homogeneous solution.

Keeping in mind that the non-unitary interactions substantiate an inter-level feedback for open systems as considered in Chapter 4, we hope that the present considerations about their role as "homogenizers" in closed systems give credible basis for the idea that the Second Law could be re-formulated so that to be viewed as a general ban over perpetuum mobile. As we will demonstrate in the next part of the book, such ban does not allow not only an unrestrained drainage of the environment but it does not allow generation of information without generation of certain amount of "noise" as well.

SELF-ORGANIZED CRITICALITY

The idea that the complex systems should be considered as systems that are not at thermodynamical equilibrium has been adopted by the scientific community long ago. Yet, the following 3 facts and observations have been chosen as decisive for the behavior of complex systems:

$1/f$ dependence of the power spectrum shape; power law distributions; and the lack of fine-tuning for any of these properties. Actually, it has been established that power law dependences are generic for the so called critical phenomena. A system is "critical" if it is in transition between two phases; for example, water with a temperature set at exactly its freezing point is a critical system. Thereby, the behavior of complex systems has been classified as a special kind of critical phenomena.

A wide variety of critical systems demonstrate a few common behaviors:

- Long-tailed distributions of some physical quantities: for example, in freezing water the distribution of crystal sizes is characterized by a power law.
- Fractal geometries: freezing water tends to form fractal patterns—the canonical example is a snowflake. Fractals are characterized by self-similarity; that is, parts of the pattern resemble scaled copies of the while.
- Variations in time that exhibit pink noise: what we call "noise" is a time series with many frequency components. In "white" noise, all of the components have equal power. In "pink" noise, low-frequency components have more power than high-frequency components. Specifically, the power at frequency f is proportional to $1/f$ visible light with this power spectrum appears pink, hence the name.

Critical systems are usually unstable. For example, to keep water in a partially frozen state requires active control of the temperature, action known as fine-tuning. If the system is left near the critical temperature, any small deviation will tend to move the system into one phase or the other.

Many natural systems exhibit behaviors indicative of criticality, but if critical points are unstable, they should not be common in nature. This is the puzzle Bak, Tang and Wiesenfeld addressed in their seminal work (Bak et al, 1987). Their solution is called self-organized criticality (SOC), where "self-organized" means that from any initial condition, the system tends to move toward a critical state and stay there, without external control.

As an example, they propose a model of a sand pile. The model is not a very realistic description of a real sand pile, but it has become the standard example of self-organized criticality.

The model is a $2d$ cellular automaton where the state of each cell, $z(i, j)$, represents the slope of a part of a sand pile. During each time step, each cell is checked to see whether it exceeds some critical value, K. If so, an "avalanche" occurs that dissipates the local slope; specifically, $z(i, j)$ is decreased by 4, and each of the 4 neighbors is increased by 1.

At the perimeter of the grid, all cells are kept at $z = 0$, so the excess spills over the edge. To initialize the system, Bak et al. start with all $z > K$ and evolve the system until it stabilizes. Then they observe the effect of small perturbations; they choose a cell at random, increment its value by 1, and evolve the system, again, until it stabilizes.

For each perturbation, they measure D, the total number of cells that are affected by the resulting avalanche. Most of the time, D is very small, either 0 or 1. But occasionally there is a large avalanche that affects a substantial fraction of the grid. The distribution of D turns out to be long-tailed, which supports the claim that the system is in a critical state.

A sand pile exhibits punctuated equilibrium behavior, where periods of stasis are interrupted by intermittent sand slides. The sand slides, or avalanches, are caused by a domino effect, in which a single grain pushes one or more other grains and causes them to topple. In turn, those grains of sand may interact with other grains in a chain reaction. Large avalanches, not gradual change, make the link between the quantitative and qualitative behavior, and form the basis for emergent phenomena.

It should be stressed once again on the difference between the critical phenomena and the SOC. In physics, a critical point is a point at which a system changes radically its behavior or structure for instance, from solid to liquid. In standard critical phenomena, there is a control parameter which a person can vary so that to obtain this radical change in behavior. In the case of melting, the control parameter is temperature. Self-organized critical phenomena, by contrast, is exhibited by driven systems which reach a critical state by their intrinsic dynamics, independently of the value of any control parameter. The archetype of a self-organized critical system is a sand pile. Sand is slowly dropping onto a surface, forming a pile. As the pile grows, avalanches that occur carry sand from the top to the bottom of the pile. At least in model systems, the slope of the pile becomes independent of the rate at which the system is driven by dropping sand. This is the (self-organized) critical slope.

Further, Bak and Sneppen (Bak and Sneppen, 1993) developed the idea of self-organized criticality by means of involving heterogeneous barrier for triggering avalanches. In turn this extents the availability of the model making it appropriate for biological modeling in accordance with the ideas of S. Kauffman (Kauffman, 1995). Next we present in brief the Bak-Sneppen model applied both to a seismic event and a model for biological evolution. The Bak-Sneppen model is a one-dimensional array of N sites. Each site represents a species, and is assigned initially a random barrier, B_i, between 0 and 1. The barrier is a measure of the fitness or survivability of the species. At each time step, the site with the lowest barrier is identified and the species at that site is mutated or assigned with a new random number. All species interact with each other

through a food chain. The interaction is introduced by assigning new random numbers to the two nearest neighbor sites of the mutated site. This procedure is repeated. Irrespective of the initial distribution, the system reaches a stationary state, displaying the self-organized critical behavior. The model can easily be explained in seismological terms, too. The fitness landscape of the BS model is equivalent to the heterogeneous barrier distribution over a fault plane generating earthquakes. Mutation corresponds to rupture. In seismology, non-uniform distribution of strength over a fault plane is called "barriers" or "asperities, "and is considered to cause the complex ruptures process of earthquakes. Irregularity is either geometrical or in the physical and mechanical properties. We can consider the barrier in the BS model as the magnitude of strength relative to tectonic stress applied on a fault in the following earthquake fault model. An earthquake, that is an avalanche of rupture, is defined to stop when the minimum barrier becomes stronger than the threshold. Another earthquake will start from the site with the minimum barrier after some time when the tectonic stress is increased again. This procedure is essentially the same as that of the biological evolution model.

Outlining, the SOC model links together these factors: a simple cellular automaton produces several characteristic features observed in natural complexity (fractal geometry, $1/f$ noise and power laws) in a way that could be linked to critical-point phenomena. Crucially, however, the SOC model demonstrates that the complexity observed emerges in a robust manner that does not depend on finely tuned details of the system: variable parameters in the model could be changed widely without affecting the emergence of critical behavior (hence, *self-organized* criticality). Thus, the key result of the SOC model is its discovery of a mechanism by which the emergence of complexity from simple local interactions could be *spontaneous*—and therefore plausible as a source of natural complexity — rather than something that is only possible in the lab (or lab computer) where it is possible to tune control parameters to precise values.

The model of self-organized criticality has an enormous impact on the development of scientific ideas in the latest decades as being the first example which demonstrates that the complexity appears as spontaneous self-organization so that the emergent phenomena are not only insensitive to the details of the microscopic local rules from which they emerge but also their self-organization does not need fine-tuning.

Yet, despite its enormous ingenuity and importance the SOC model suffers serious drawbacks:

1. It cannot mathematically prove that systems follow the power law, i.e. the SOC model does not provide a criterion for discrimination between systems that exerts power law or not.
2. Most changes occur through catastrophic events rather than through gradual changes. This happens because once the sand pile model reaches its critical state there is no correlation between the system's response to a perturbation and the details of a perturbation. Generally this means that dropping another grain of sand onto the pile may cause nothing to happen, or it may cause the entire pile to collapse in a massive slide.

The SOC theory is an abstract statistical parameter-free theory aimed to explain out-of-equilibrium phenomena. Yet, its fundamental drawback is that it leaves the response unrelated to the details of the perturbation and thus it cannot incorporate the specificity of a system.

We view the concept of boundedness as further development of the SOC model. Indeed, we adopt the idea about the thresholds that once exceeded the system would become unstable. Yet, there are essential differences. The first one is that even though we consider the complex systems as systems that are not in the state of traditional equilibrium, we do not associate their behavior with any form of criticality. On the contrary, we build our approach on the basis of assuming stability of the complex systems. The second fundamental difference between the concept of boundedness and the SOC model is that we put forward the idea that the response is local and specific and is determined only by the current state and the current impact; the only constraint we impose is that of boundedness thereby, we open the door to "critical"-type phenomena substantiated through gradual changes. The next fundamental difference between the SOC model and the concept of boundedness is that while the SOC model does not imply any hierarchical order, according to the concept of boundedness the power dependences are due to a great extent to the role of the hierarchical organization. Last but not least, the SOC model cannot give rise to coexistence of a discrete and a continuous band in the power spectra; thus, the SOC model is incompatible with any form of "homeostatic" type of response.

DETERMINISTIC CHAOS: MAJOR PROBLEMS

Deterministic chaos is a mathematical theory which studies the origin of the well-established group of properties of complex systems known as sensitivity to initial conditions. These phenomena imply the following:

1. Long term behavior is difficult or impossible to predict: Even very accurate measurements of the current state of a chaotic system become useless indicators of where the system will be. One has to measure the system again to find out where it is.
2. Sensitive dependence on initial conditions: starting from very close initial conditions a chaotic system very rapidly moves to different states.

3. Broadband frequency spectrum: that is, the output from a chaotic system sounds "noisy" to the ear. Many frequencies are excited.
4. Local instability versus global stability: in order to have amplification of small errors and noise, the behavior must be locally unstable: over short times nearby states move away from each other. But for the system to consistently produce stable behavior, over long times the set of behaviors must fall back into itself.

Therefore, the major problem that the above observations pose is why the chaotic processes basically work as amplifiers: they turn small causes into large effects. That means that small, unobservable fluctuations will affect the outcome of the process. Although the process is deterministic in principle, (which implies equal causes having equal effects), it is unpredictable in practice. One of the widespread views on this puzzle is that causes that seem equal to the best of our knowledge can still have unobservable differences and therefore lead to very different effects. Yet, this view faces the question why we are able to establish deterministic rules with finniest precision along with always finite inaccuracy of establishing initial conditions.

Another puzzle posed by the chaotic properties commences from the interplay between the local instability of the behavior and its global stability. It should be reminded that this interplay does not exactly match the sensitivity to initial conditions. This mismatch comes out from the fact that their coexistence is possible only for non-linear systems; but for linear systems it is impossible to have both sensitivity to initial conditions and "stretching and folding", that is the mechanism that brings about the coexistence of local instability and global stability. To explain this let us present the following example:

$$x_{n+1} = 2x_n \tag{6.3}$$

The distance between two different solutions increases by the factor two in each time step. But of course, this is not deterministic chaos! It is a kind of explosion (like the population explosion!). Sensitivity on the initial conditions leads to chaos only if the trajectories are bound. That is, the system cannot blow up to infinity. With linear dynamics, you can have either sensitivity on the initial conditions or bound trajectories, but not both. With nonlinearities, you can have both. Thus, *stretching and folding* are responsible for deterministic chaos. And there is no folding without nonlinearities! This makes the mechanism of stretching and folding hallmark for deterministic chaos. Therefore, the quest for deterministic chaos turns into quest for those non-linearities which brings about stretching and folding.

Thereby, the phenomenon of deterministic chaos appears as complementary to the self-organized criticality: indeed, while for the self-organized criticality the properties related to criticality are not sensitive to the details of local dynamical rules, the existence of stretching and folding is particularly sensitive to the details of the local rules. Then, we face the fundamental dilemma whether both "critical"-like properties and the stretching and folding are counterparts in a new concept that gives rise to their self-consistent coexistence. A crucial test for such concept would be the explanation of each of these properties.

A concept that accomplishes this task is the concept of boundedness: it gives rise both to "critical-like power dependences and the properties of the deterministic chaos. Moreover, it answers successfully the following two questions of fundamental importance: (i) since the chaotic properties are established at a wide variety of natural systems, is there a general enough concept that is able to associate the evolution of that broad spectrum of natural systems with single mathematical structure - that of the systems of differential equations; if so, how the Nature drives the irregularities of the chaotic time series in every particular system; (ii) though the properties of the chaotic solutions are well established now, the relation between the determinism of the differential equations and the statistical properties of their chaotic solutions is still open; moreover, the puzzle increases because it has been found out that the chaotic properties are insensitive to the statistics of the chaotic time series. Hence, we face a crucial question: what type of statistics brings about the insensitivity of chaotic properties and how it is related to the determinism of the differential equations. Going a step further, this question opens the door to challenging the very idea of deterministic chaos, namely: focusing attention on the type of statistics that provides properties insensitive to its particularities makes the matter about the mathematical origin of the irregularities irrelevant. In other words, we put the idea of deterministic chaos upside down: we ask whether there is a general concept that renders some properties of the natural time series independent of their statistics. This paraphrase is not only of particular interest for the mathematicians but it has deep epistemological understanding as well. Indeed, our knowledge about the Nature is built on the idea that we are able to extract a genuine for the system signal from the "noise" that inevitably accompanies its recording. Furthermore, the non-linearities of the recording due to effects like local averaging, inertia, resolution, filtering etc. distort the genuine signal. However, the distortion is crucial for an irregular, "noisy" signal, chaotic ones included, because it "transforms" the original time series into a new one. This immediately illuminates the importance of the issue about the existence of such properties of the time series that are independent of their statistics. We assert that such general concept does exist and it is the concept of the boundedness. Furthermore, in Chapter 1 we have proved that time series subject to that concept exhibit properties insensitive to the particularities of their statistics – to the most surprise, those properties are identical to the corresponding ones of the chaotic time series. Now we are going to do more: in section

Simulated Dynamical Systems under Boundedness of the present Chapter we will demonstrate that the numerical simulation of chaotic dynamical systems appears in the frame of the concept of boundedness. Yet, prior to that we will briefly illustrate how the major chaotic properties commence from the concept of boundedness.

CHAOTIC PROPERTIES UNDER BOUNDEDNESS

The task of the present section is to illustrate how the chaotic properties emerge from the standpoint of the concept of boundedness. Let us first focus our attention how the concept of the boundedness is incorporated in the structure of the time series. Obviously, that boundedness imposes two general constraints on every time series: the first one is that the amplitude of its terms is bounded so that the margins of their variations are dictated by the thresholds of stability; the second one is that the boundedness of the rate of development of the fluctuations renders interrelation between the size of a fluctuation and its duration. Now we must prove that these constraints are enough to ensure the target insensitivity of the chaotic properties from the statistics of a zero-mean time series subject to boundedness; we call hereafter such time series BIS (bounded irregular sequence).

We start with a very important characteristic of every irregular sequence, namely its autocorrelation function whose definition reads:

$$G(\eta) = \lim_{T \to \infty} \frac{1}{T} \int_0^{T-\eta} \left(X(t+\eta) - \langle x \rangle \right) \left(X(t) - \langle x \rangle \right) dt$$

(6.4)

The autocorrelation function is a measure of the average correlation between any two points in the sequence separated by a time interval $\eta \leq T$. An intriguing interplay of the boundedness and the scaling invariance is developed. On the one hand, the uniform contribution of all time scales restrains persistence extent without signaling out any specific one. On the other hand, the boundedness limitates the persistence because every deviation inevitably "turns back". Note that the uniform contribution of all time scales is very different from random contributions. The latter is characterized by the lack of any systematic correlations whilst the former is characterized by persistent but parameter-free correlations. But how far away is the persistence spread? And how the boundedness affects its extent? In section *Autocorrelation Function* of Chapter 1 we found out that the autocorrelation function of BIS of arbitrary but finite length T reads:

$$G(x) = \sigma \left(1 - (x)^{\nu(x)} \right) \qquad (6.5)$$

where $x = \dfrac{\eta}{T}$, $x \in [0,1]$; σ is the variance of the BIS and $\nu(x) = 1 - x$. It is obvious that the shape of $G(x)$ meets the requirement that the autocorrelation function is parameter-free. It should be stressed that the derivation of the above shape involves only constraints imposed by the boundedness. In turn, this immediately implies that the shape of the autocorrelation function is insensitive to the statistics of the time series.

Another important characteristic of every time series is its power spectrum. Though it is Fourier transformation of the autocorrelation function, it has its own role among the properties of the time series. It turns out that the shape of the power spectrum of every BIS is also insensitive to its statistics and uniformly fits the shape:

$$S(f) \propto 1 \big/ f^{\alpha(f)} \text{ for } f \geq \frac{1}{T} \qquad (6.6a)$$

$$S(f) \equiv 0 \text{ for } f < \frac{1}{T} \qquad (6.6b)$$

where $\alpha(f) = 1 + \kappa \left(f - \dfrac{1}{T} \right)$; T is the length of the time series and κ is a parameter set on the thresholds of stability.

The insensitivity of the shape of the power spectrum to the statistics of the time series opens the door to separation of genuine for the system information from the inevitable internal and external noises and their distortion by the recording. Note that the long-range correlations due to a genuine for the system process appear as discrete lines (bands) in the power spectra. Thus, the insensitivity of the "noise" power spectrum shape from the statistics lets unambiguous and reproducible separation of the discrete bands (genuine signal) from whatever noise under the mere condition that the latter is to be bounded.

So far we have established that the power spectrum and the autocorrelation function of the BIS exhibit remarkable universality - their shape and characteristics are free from any specification of the fluctuation dynamics. Though this ubiquitous insensitivity has been rigorously proven, our intuition remains unsatisfied and we are strongly tempted to suggest that the power spectrum and the autocorrelation function are too "coarsen" characteristics to take into account the particularities of the fluctuations dynamics. Indeed, the dynamics of the fluctuations comes out from specific for every system processes associated with their development. Now we will show that the fluctuation dynamics is also subject to certain universality.

The usual way to reveal the dynamics of the correlations in a stochastic sequence is to study the properties of its phase space. The procedure involves dividing of the phase space volume into small size cells and counting the points at which the trajectory intersects each of them. It is certain that this elaborate operation helps much in the examination of the correlation dynamics. Alongside, it is to be expected that the obtained information is highly specific. However, the general aim of this section is to illuminate the universal properties of the BIS and their relation to the chaoticity. Our major goal is the systematic derivation of all chaotic properties from the boundedness. That is why now we focus the attention on the question whether the boundedness gives rise to some universality of the fluctuation dynamics. In section *Boundedness of the Rates. Embedding Dimension* in Chapter 1 we have proved that the interrelation between the size and the duration of a fluctuation drives a mechanism of stretching and folding in the phase space. This along with the presence of thresholds of stability makes the fine structure of the attractor very different from the coarse-grained one. The correlation dimension of the attractor vs. cell size at 3 different topological dimensions is plotted in Equation 1.2. Though we do not specify the actual topological dimension n, the plot is made so that to leave the dynamical rate intact. The plot exhibits the following characteristic properties of the population behavior:

1. At every topological dimension the plot manifests nearly linear part. Accordingly the population behaves as l^{-D_n} where D_n decreases on n increasing.
2. At every topological dimension the population reaches saturation;
3. The saturation value is insensitive to the value of the topological dimension.

The linear part of the plot is apparently associated with the prevailing role of the stretching. Further, since the folding shrinks the distance between the points, it effectively weakens the correlations necessary for stretching. In other words it acts towards stochastisation of the phase space points. Thus the population tends to remain constant and insensitive to any further growth of the cell size. This immediately renders the saturation value insensitive to the topological dimension.

It is worth noting that the above plot is pronounced illumination in favor of the boundedness concept because the above behavior of the correlation dimension is one of the most pronounced chaotic properties. Furthermore, it is a not only a property related to the chaos but it turns out to be insensitive to the statistics of the time series as well. So, we again experience the remarkable universality that the boundedness brings about.

Another highly specific for the deterministic chaos characteristics is the so called K-entropy. It is a measure of the average information necessary for precise setting of the motion in the phase space. The value of K-entropy is definitive for the chaos since it is zero for the deterministic motion, infinite for the stochastic one and finite for the deterministic chaos. So, what is its value for a BIS? In section *K-entropy* of Chapter 1 we established that the behavior of K-entropy commences straightforwardly from the mechanism of stretching and folding. The explication of that commence becomes evident by setting $K(u)$ through $(1 - P(u))$, where $P(u)$ is the probability for participation to the stretching and folding; u is the phase space angle. Then the plot of $K(u)$ vs. u presented in Equation 1.3 has the following meaning: the increase of the probability for stretching at small u requires more information for setting the distance between the initial and the final point of any trajectory. Further, the stretching reaches its maximum and turns to folding which, however, shrinks the distance between the initial and the final point of the trajectory. Thus the folding contributes to better precision of the position of the motion. In turn, the better precision needs less information for specifying the characteristics of the trajectory. As a result, the K-entropy turns to decrease.

A decisive property of K-entropy is that it presents the stretching and folding in power type form. Let us outline that the power type form is the only possible parameter-free form that meets simultaneously the following two requirements: lack of entanglement among time scales and boundedness of transmitting substance trough space-time (dynamical boundedness).

Summarizing, let us mention once again that all the properties of BIS presented above are derived without involving any information about the statistics of the time series. At the same time, however, the derivation explicitly exploits the idea of boundedness and the time-translational invariance. These circumstances let us to conclude that time series subject to boundedness concept manifest major chaotic properties regardless to the origin of the fluctuations. In turn, the latter gives priority to the boundedness concept as the onset of the chaocity since it is not anymore related to a single mathematical structure (deterministic differential equations). Yet, the crucial test for the boundedness concept is whether the deterministic chaos that appears from the simulation of some dynamical systems can be put in the frame of the boundedness concept. If so, this result will justify an undisputable superiority and will ensure an enormous ubiquity of the boundedness concept. We will discuss this issue in the next section.

SIMULATED DYNAMICAL SYSTEMS UNDER BOUNDEDNESS

The low dimensional chaos is associated with the dynamical systems of the type:

$$\frac{d\vec{x}}{dt} = \vec{\alpha}\hat{A}\left(\vec{x}\right) - \vec{\beta}\hat{R}\left(\vec{x}\right) \qquad (6.7)$$

where $\vec{\alpha}$ and $\vec{\beta}$ are control parameters. Equation (6.7) are a system of ordinary differential equations that fulfill Lipshitz conditions. Then its solution should be smooth line whose course is predetermined by the initial conditions in an arbitrarily long span both back and forward in the time. However, it turns out that numerical solution

of some dynamical systems behaves differently at certain values of the control parameters $\vec{\alpha}$ and $\vec{\beta}$: it is highly irregular and exhibits strong sensitivity to the initial conditions. It has been established that this behavior arises around unstable solution(s). But what drives the solution to behave rather as a solution of stochastic than of deterministic equations? It is to the point to mention that the name 'deterministic chaos' comes namely from the controversial issues posed by following vicious circle: though Equation (6.7) are deterministic, i.e. they do not involve any stochastic functions, their solution behaves as a stochastic function that exhibits chaotic properties. It has been established that its power spectrum is a continuous band; its embedding dimension is finite and fractal. However, the great mystery remains: why the solution behaves as an irregular function?

To reveal the mystery let us remind that the numerical simulation involves inevitable round-off at every step. When the solution is stable the round-off is negligible. However, at unstable solutions the round-off is amplified and rapidly becomes compatible to the values of the smooth part. Further, it appears as multi-valued function since the round-off around 0.5 is executed as a random choice among two selections: the round-off to the lower value and the round-off to the higher value. Then, the amplification of the round-off causes "transformation" of the deterministic Equations (6.7) into the following stochastic ones:

$$\frac{d\vec{x}}{dt} = \vec{\alpha}\hat{A}\left(\vec{x}\right) - \vec{\beta}\hat{R}\left(\vec{x}\right) + \vec{\alpha}\hat{\eta}_{ai}\left(\vec{x}\right) + \vec{\beta}\hat{\eta}_{rj}\left(\vec{x}\right)$$

$$(6.8)$$

where $\hat{\eta}_{ai}\left(\vec{x}\right)$ and $\hat{\eta}_{rj}\left(\vec{x}\right)$ are the stochastic terms that come from the amplification of the round-off. Moreover, the stochastic terms in (6.8) are bounded since it is well established that the solution of the chaotic dynamical systems is confined to the so called strange attractor that has finite volume. Thus, the solution of a simulated dynamical system (6.8) is BIS unlike the solution of (6.7) that is smooth function the course of which is predetermined by the initial conditions.

Note that the stochastic terms in (6.8) are pronounced only at those values of the control parameters where the solution of Equation (6.7) is unstable. Thus, the chaoticity is manifested only at particular control parameters choice. It should be stressed that the association of the chaoticity with specific control parameters is a property generic only for the simulated dynamical systems. Unlikely, the manifestation of the chaoticity in the natural systems is free from any general specification of the control parameter choice since the boundedness provides independence of the chaoticity from origin of the fluctuations and from the mathematical structure that describes them. Moreover, under boundedness, the chaoticity is to be associated with stable solutions only; thereby, the global stability of the solutions of the "homeostatic" part (presented by equations of the type (6.7)) interplays with the local instability of the "stochastic" part of the solution thus providing the mechanism of stretching and folding.

Summarizing, the boundedness concept manages to explain in a very natural way the greatest mystery of the deterministic chaos: the origin of the irregularities of the solution. The key point is that in the domain of unstable solutions the simulation transforms the deterministic equations into stochastic ones whose solutions are BIS.

CONCLUSION

The behavior of complex systems is one of the greatest challenges to the modern interdisciplinary science. It encompasses a huge diversity of systems which share the same behavior: open flow chemical systems, earthquakes, weather, colonies of ants, telecommunications, DNA sequences, and human languages to mention a few. The main line that remains stable is that despite the ubiquity and wide diversity of physical and social phenomena all those systems share certain universal properties preserving individuality in each case. Thus, the major questions we face is why is it necessary for all those systems to share certain universal properties; how the systems control their behavior so that to be able both to exert universality and preserve their individuality? The puzzle increases enormously since the universal behavior cannot be explained in the frame of the traditional physical science. Indeed, properties like ubiquitously observed power dependences are rather typical for the critical phenomena of the phase transitions. However, while the phase transitions need fine-tuning to occur, the power dependences appear at every value of the control parameters. Thus, even if we adopt the idea that the power dependences are expression of certain form of criticality, does it imply that the Universe lives at the edge of criticality? This point of view is also undermined by the fact that the power dependences are in confrontation with the time-translational invariance of the observed phenomena. The presence of power dependence implies that the exchanged matter/energy in any time interval depends on the length of the time window; in addition, the size of the most probable fluctuations also increases with the length of the window. Therefore, the behavior of a system is not time-translational invariant and should be highly unstable. However, no collapse of a system displaying power dependences happens; on the contrary, it stays stable long before and long after any observation.

Then, we face the question whether reading of the empirical data is correct. This question arises always when the existent theory fails to explain certain facts and when certain facts cannot be incorporated self-consistently into the frame of the existent theory. The problem with the behavior of the complex systems concerns the entire paradigm of the traditional reductionist approach to the matter organization: the question what makes a system to exert permanent fluctuations and at the same time stay stable suggests that the system regulates effectively the characteristics of the fluctuations. But how does it happen: appearing at macroscopic level, these fluctuations must be a result of coherent behavior at dynamical level. Since at atomic level the constituents exert free motion between collisions, the fluctuations that they create are to be independent from one another and hence to average out. Then, how is it possible a complex system to behave in an irregular manner? If we assume the idea about certain long-distance coherence, how the latter is implemented by means of the short-range interactions between atoms and molecules. Furthermore, even if there would be certain synchronization; its relaxation time would be proportional to the size of the system. Thus, the bigger the system, the slower its "reaction" would be. The consequences of the above considerations outline a contradictory situation: if a system is hierarchically organized, the higher levels would relax faster then the lower ones because the higher level has fewer components than its predecessor. This poses the question about the "hierarchical levels" interactions: do they go only bottom up as the traditional reductionist approach assumes, or it is to be expected that a hierarchy is organized as feedbacks that operate in both directions, i.e. bottom up and top down?

The theoretical and practical grounding for the suggestion that the hierarchical level "interactions" operate as feedbacks in both directions (bottom up and top down) is straightforwardly related to the question about the impact of one level onto another. In other words, it is the question whether the impact is maintained bounded. The reason for this question comes from the well-known 'gedanken' phenomenon "butterfly effect". It has been coined by the meteorologist Lorentz who, while studying the equations that determine the weather, noticed that their outcomes are strongly dependent on the initial conditions. The weather is a chaotic system. The tiniest fluctuations in air pressure in one part of the globe may have the most spectacular effects in another part. Thus, a butterfly flapping its wings somewhere

in Chicago may cause a tornado in Tokyo. It becomes a dominant view that this effect explains why scientists find it so difficult to predict the weather. To predict future situations, they need to know the present situation in its finest details. But obviously they will never be able to know all the details: they cannot monitor every butterfly flapping its wings! The fewer details they know, the less accurate their long term predictions will be. That is why reliable weather predictions can be made for no more than a few days in advance.

The considerations about the size of the "butterfly effect" pose the question about the relation between the largest size of a fluctuation and the size of the corresponding system. The traditional statistical physics asserts that the largest fluctuations are proportional to the size of the system (to square root actually). An immediate consequence of that approach is that the larger a system, the more pronounced "butterfly effect" is. If correct, this result would render any astronomical observation false since the tiniest errors would be "amplified" to enormous extent by the distances. In turn, if not restrained, the "butterfly effect" would make any form of epistemology impossible; it would render life impossible and it would render the Universe unstable. The way out from this tricky problematics goes through supposition that both the impact and the response are to be bounded at every level of matter organization. In turn, the suggested boundedness brings about the stretching and folding mechanism which guarantees that the fluctuations are self-organized so that to exert U-turns at the thresholds of stability. However, the assumption about the independence of the size of fluctuations from the size of the system requires an entirely new approach to the subject-matter.

The above considerations delineate the fundamental differences between the existing approaches and the concept of boundedness:

The focus of the concept of boundedness is concentrated on the long-term stability of the complex systems. Then, their response to any tiniest environmental impact is supposed local, i.e. specific for the state and for the current impact, while remaining active, i.e. self-organized so that to be bounded and thus to keep the system within the margins dictated by its thresholds of stability. Thus, it is expected that the changes in the system happen gradually. On the contrary, the rival theories, such as SOC and its complementary deterministic chaos, do not specify any relation between the impact and the response: moreover the lack of its specification turns out to substantiate the properties they are aimed to explain. It is worth noting, that the concept of boundedness successfully explains both the power dependences and the "sensitivity" to initial conditions. The greatest merit of that explanation is that the universal properties of the complex systems turn out to be remarkably insensitive to the details of that specification. They appear as time series invariants whose generic property is their invariance to the statistics of the time series;

1. Neither the model of self-organized criticality nor the chaos theory gives coexistence of specific and universal properties. Moreover, they "transform" the specific properties (viewed as the local rules which govern the dynamics) into universal ones only. On the contrary, the concept of boundedness gives rise to coexistence of specific and universal properties as a generic property of the "homeostatic"-type of response (viewed as coexistence of a discrete and a continuous band in the power spectra).

2. The replacement of the idea about thermodynamical equilibrium as the only state at which a system could reside arbitrary long-time with the idea of homeostasis opens the door to incorporation of intelligent systems into already delimited class of natural phenomena and sets credible ground for the view that

the response of a complex system could be self-organized in a semantic-like manner.

3. The value of the SOC theory and the deterministic chaos is that they introduce the very important notions of thresholds of stability and of mechanism of stretching and folding correspondingly. We admit that we put these notions at the very base that substantiates the developed by us concept of boundedness. However, as it becomes clear from the considerations in the present Chapter, put in a context different from the original one, these notions give rise to fundamentally different results. Thus, we demonstrate not only the merits of our approach, but to focus our attention on the methodological importance of the general frame through which we view a wide variety of empirical observations. It is a rare occasion in the science when the choice of general view and logic turns out so crucial for substantiating an approach with so far going consequences.

REFERENCES

Bak, P., & Sneppen, K. (1993). Punctuated equilibrium and criticality in a simple model of evolution. *Physical Review Letters*, *71*(24), 4083–4086. doi:10.1103/PhysRevLett.71.4083

Bak, P., Tang, C., & Wiesenfeld, K. (1987). Self-organized criticality: An explanation of 1/f noise. *Physical Review Letters*, *59*(4), 381–384. doi:10.1103/PhysRevLett.59.381

Kauffman, S. (1995). *At home in the universe: The search for laws of self-organization and complexity*. New York, NY: Oxford University Press.

Section 2
Boundedness:
Properties of Self-Organized Semantics

INTRODUCTION

Boundedness: Properties of Self-Organized Semantics

The major goal of the present book is to establish what makes a complex system "intelligent" and why it should share certain "indifferent" to the intelligence properties. We started revealing this problem from the second part of this question: in section 1 of the book we have considered the typical for complex systems properties regardless to whether they are intelligent or not. The novel theoretical approach manages to explain the link between two empirical facts: sensitivity to initial conditions and power laws that have been considered so far as independent from one another. What constitutes the merit of our approach is that we derive these properties from the stability of a system. Thus, the main advantage of this new theory is set by its exclusive ability to reconcile the confrontation between the power laws and sensitivity to initial conditions on the one hand and the empirically observed stability of the complex systems on the other. Yet, we expect more from our approach: we expect that the concept of boundedness which we put forward as an operational protocol for governing the behavior of complex systems also governs the intelligent-like response of a large class of complex systems.

Let us start with specifying the notion of an intelligent-like response. By this we understand an active response which is not reducible to a sequence that exactly matches the properties of the corresponding impact. To compare, the traditional approach of statistical physics sets the response passive, i.e. it always follows the environmental impact. Instead, our concept of boundedness suggests the response to be determined by local factors only: it is specific and is determined by current local state and current local impact. Yet, the response and the impact are subject to the mild constraint of boundedness which asserts that the response is self-organized so that to stay within margins dictated by the thresholds of stability and so that the rate of exchanging matter/energy with the environment is kept bounded. Then, the response turns out to be a two-component variable: an active specific self-organized pattern which is robust to small environmental fluctuations and a passive, universal component which retains certain characteristics which are insensitive to the statistics of the impacts and responses. That is why we called the robust self-organized part of the response "homeostasis" and the other non-specific one, noise component. A systematic study of the generic properties of the coexistence of the homeostasis and noise is presented in the first part of the book.

A decisive for our concept of boundedness ingredient is the assumption that the strengthening of the response goes through hierarchical self-organization. Chapter 4 proved that the substantiation of the hierarchical functioning through the basic inter-level feedback yields strengthening of the response. The turning point of this approach is that the strengthening of the response is executed in a semantic-like manner.

Even though the established minimal for a semantic-like response conditions are quite mild, we pose the question how typical is their upgrade so that it to be able to develop writing "words" into writing "sentences." We must start answering this question with establishing the general properties of the state space which a system exerting complex semantics has to retain. First, it is obvious, that to make a "sentence" out of "words" it is necessary that the words must be viewed as different units according to their role in a "sentence." Put in other words, the meaning of a word in a sentence is irreducible to the sequence of letters it is constituted of. In order to achieve this task each word must be characterized by certain metric-dependent property so that we to be able to define"distance" between words. In order to make this measure typical for a sentence, it must be subordinated to a general rule available along the entire sentence. Thus, we are very close to the idea that a "sentence" is characterized by a variable whose meaning is close to that of the physical action. Yet, since complex systems are open, the "action" could not be derived by a variational principle as in the traditional physics. Then, how to define it? One of the major merits of the action, defined in Chapter 7 is that it implements the hierarchical order in the direction 'top down' and thus substantiates a complex interplay of the hierarchical order that goes in the directions 'bottom up' and 'top down' at the same time. In turn, this interplay opens the door to an increase of the diversity of the semantic-like response by means of its non-extensive hierarchical super-structuring.

Another important issue is the one that concerns the energy and entropy costs of the information. A question arises about the relation between semantic-like response organized into non-mechanical engines and the long-standing problem about the existence of *perpetuum mobile*. The issue about the possibility for existence of *perpetuum mobile* is subject to the Second Law which nowadays is considered as one of the fundamental laws of Nature. Along with the question concerning *perpetuum mobile*, the Second Law is assumed to reveal one of the principal early mysteries of thermodynamics, namely the origin of irreversibility: while microscopic equations of motion describe behaviors that are the same in both time directions, why do large-scale systems exhibit temporal asymmetries? Many thermodynamic processes go in one direction: closed systems devolve from order to disorder, heat flows from high temperature to low temperature, and shattered glass does not reassemble itself spontaneously. These are described as transient relaxation processes in which a system moves from one macroscopic state to another with high probability since there is an overwhelming number of microscopic configurations that realize the eventual state. The Second Law asserts that the natural processes spontaneously evolve in the direction of increasing entropy so that eventually to reach equilibrium, the state of maximum entropy, where a system stays arbitrarily long time.

The relation between semantics and the arrow of time is grounded on the understanding of sensitivity to permutations of a given semantic sequence (unit). It is well known that each semantic unit is sensitive to permutations of its constituents: permutation sensitivity is an implement of distinguishing meaningful sequences of letters from the random ones. Therefore, the semantic meaning selects an "arrow" of meaning which is typical for each semantic unit. The problem is highly non-trivial since the "arrow" of meaning is a complex interplay between linear and circular behavior. Indeed, while inside each semantic unit this arrow is linear, on accomplishing the unit, i.e. when the unit returns to the "space bar," the "arrow" turns circular. At this point, it is worth reminding that the semantic units on each hierarchical level of semantic organization are separated by punctuation marks: the words are separated by "space bars," the sentences are separated by dots, the paragraphs are separated by a blank line, etc. Thus, it seems that semantics is in conflict with the Second Law since it does not select a monotonic arrow of time. This, however, poses the question whether it is ever possible to have a semantic-like response of natural systems? In this line of considerations it should be pointed out that a Turing machine also does not select a monotonic "arrow of meaning." Indeed, since it is *a priori* impossible to decide which algorithm halts, it is never known whether a given algorithm ends in "cycling" or accomplishes its task. Thus it is impossible to decide whether the algorithm selects a monotonic "arrow of meaning" or it runs an endless repetition of a "cycle."

We come to a point where we face the dilemma whether the information processes are an exception from the Second Law or the Second Law calls for certain reformulation which is equally available for simple physical systems and for the complex systems exhibiting semantic-like response. The association of permutation sensitivity with the irreversible performance of a non-mechanical 'engine' inevitably bounds the issue about the arrow of time with the issue about the *perpetuum mobile*. Chapter 9 defines the Second Law viewed as ban over *perpetuum mobile* avoiding the condition that the equilibrium is the state of maximum entropy. Alongside we define the arrow of time in a completely new way. In turn this makes the Second Law ubiquitous: now it is available not only for simple physical systems but it encompasses the complex system such as patterns, living organisms etc. as well.

Last, but not least, Chapter 10 considers the new understanding which certain aspects of the algorithmic theory acquire in view of presented by us approach. Thus, for example the notion of causality acquires another "dimension" – the association of the semantic content with the irreversible performance of an 'engine' makes the notion of causality rather functional property than a "static" property related with certain time-asymmetric relations among certain variables as in the traditional algorithmic theory.

Chapter 7
Hierarchical Order

ABSTRACT

A generic property of the motion along every allowed trajectory in a semantic state space is that Hiesenberg-like relations hold. Under the condition that the action on each allowed trajectory is stationary, it is possible to define "energy," "velocity," and "mass." The so defined energy is not constant along the trajectory because of permanent mass-energy transformations. Yet, unlike Einstein mass-energy relation, the amount of energy and mass that is transformed is permanently kept bounded along the allowed trajectories. The control over local accumulation of matter/energy in a structured network is implemented through spontaneous emitting of a matter wave. The stationarity of action governs grammar rules, which sets non-random picking of successive semantic units, i.e. word order is a sentence, thus providing non-extensivity of the semantic-like hierarchy.

INTRODUCTION

The major part of our research is to establish what makes a complex system "intelligent" and to answer the question why it has to retain certain "indifferent" to intelligence properties. In the first part of the book we have considered the second part of this research issue. We have successfully explained the typical for complex systems properties regardless to whether they are intelligent or not. Thus, we manage to put in a self-consistent way the systems` sensitivity to initial conditions and the power laws. We relate these properties to the stability of a system. Thus, the main advantage of our approach is set by its exclusive ability to reconcile the confrontation between the power laws and sensitivity to initial conditions on the one hand and the empirically observed stability of the complex systems on the other. Yet, it is expected more from this approach: we expect

DOI: 10.4018/978-1-4666-2202-9.ch007

that the concept of boundedness which we put forward as an operational protocol for governing the behavior of complex systems also governs the intelligent-like response of a large class of complex systems.

At first let us specify the notion of an intelligent-like response. By this we understand an active response which is not reducible to a sequence that exactly matches the properties of the corresponding impact. To compare, the traditional approach of statistical physics sets the response passive, i.e. it always follows the environmental impact. Instead, our concept of boundedness suggests the response to be determined by local factors only: it is specific and is determined by current local state and current local impact; the response and the impact are subject to the mild constraint of boundedness which asserts that the response is self-organized so that to stay within margins dictated by the thresholds of stability so that the rate of exchanging matter/energy with the environment is kept bounded. Then, the response turns out to be a two-component variable: an active specific self-organized pattern which is robust to small environmental fluctuations and a passive, universal component which retains certain characteristics which are insensitive to the statistics of the impacts and responses. That is why we called the robust self-organized part of the response "homeostasis" and the other non-specific one, noise component. A systematic study of the generic properties of the coexistence of the homeostasis and noise is presented in the first part of the book.

A decisive for our concept of boundedness the assumption is that the strengthening of the response goes through hierarchical self-organization. In Chapter 4 we have proved that the substantiation of the hierarchical functioning through the basic inter-level feedback indeed yields strengthening of the response. The turning point is that the strengthening of the response is executed in a semantic-like manner. This can easily be understood taking into account that the homeostasis is not a single state but it has to be associated with the following generic structure of the state space: the latter is partitioned into domains (basins-of-attraction) each of which is characterized by a specific "homeostatic" state. All domains are tangent to a single point called accumulation point. Then, associating each basin-of-attraction with a letter and the accumulation point with a space bar, every trajectory in the state space "writes" "words." What makes this response active is that writing a "word" is not dictated by the environmental impact only since not every sequence of letters is consistent with the boundedness. The constraint imposed by the boundedness makes the admissible sequences of letters to go only through adjacent basins-of-attractions. Therefore, we have established the most primitive form of intelligent-like response: writing "words." We have found out that writing "words" is rather typical for the complex system behavior since the conditions for its substantiations are fulfilled by large amount of systems starting with open heterogeneous chemical systems. Yet, this semantics still comprises only writing of words not sentences. This happens because all letters are equiprobable at the accumulation point; then each word can precede and succeed any other. However, "writing" sentences requires "picking" of each specific word depending on a relating range of foregoing ones. Thereby, the accumulation point that is "space-bar" turns subordinated to a higher level super-structure of the state space. On the other hand, every sentence starts and ends by a dot. Thus, the super-structure must also be orbital so that the orbits pass trough complex basins-of-attractions each of which is constituted by a "word." The intrigue of the problem is set by the following controversy: on the one hand these considerations suggest certain self-similarity in the structure of the state space; on the other hand, the necessity of morphogenesis as a necessary minimal condition for exerting any semantics suggests that the self-similarity cannot be substantiated by straightforward application of the idea of self-organization as an operational protocol. To remind, we have proved in Chapter

4 that the self-organization viewed as an operational protocol brings about the semantic-like organization of the response. Yet, our proof has been explicitly grounded on the involvement in the equation-of-state of the inter-level feedback which maintains the local fluctuations of the concentration bounded. In view that a higher level semantics (writing sentences) is to be associated with a structured network, is it to be expected that the self-similarity of the semantic-like response is expressed as an operational protocol? We assert that there exists an abstract operational protocol which though operating in a structured network, turns out to be insensitive to the details of its structure. Its role is to outline the general road to delineating the trajectories which retain the property of responding in a semantically structured way.

Even though the minimal for a semantic-like response conditions are quite mild, we pose the question how typical is their upgrade so that to develop writing "words" into writing "sentences." We must start answering this question with establishing the general properties of the state space which a system exerting complex semantics has to retain. Hereafter we call the state space of a system whose response exhibits complex semantics a semantic state space. First, it is obvious, that to make a "sentence" out of "words" it is necessary that the words must be viewed as different units according to their role in a "sentence." Put in other words, the meaning of a word in a sentence is irreducible to the sequence of letters it is constituted of. In order to achieve this task each of them must be characterized by certain metric-dependent property so that we to be able to define "distance" between words. In order to make this measure typical for a sentence, it must be subordinated to a general rule available along the entire sentence. Thus, we are very close to the idea that a "sentence" is characterized by a variable whose meaning is close to that of the physical action. Indeed, in physics the admissible trajectories are selected by the stationarity of their

action. We shall demonstrate that in the present context, the stationarity of the action selects the trajectories which write meaningful 'sentences" as a response. What makes the present notion of action different from the traditional physical one is that it is not to be associated to any variational principle, i.e. it does not select trajectories with extreme properties such as minimal energy or maximum entropy. It should be stressed that this is impossible since the complex systems are open flow systems which permanently exchange matter/energy with the environment and thus their energy and matter permanently vary in the time course; therefore they cannot be considered as systems of conserved energy and matter; then neither variational principle is relevant for them. It should be stressed that in each elementary act of interaction the energy/matter conservation laws are fulfilled. Yet, because of the fluctuations at quantum hierarchical level, the total matter/energy of the system is not conserved; in turn this makes the variational principle irrelevant for defining our "action."

Thus we face the question how to define the notion of a stationary action avoiding the condition that the variational principles are subordinated to certain extreme properties. We will establish that it is still possible to define action but its characterization appears in a highly non-trivial way from the properties of the quantum hierarchical level set by the boundedness.

BACKGROUND

The major assertion of the axiomatic theory presented earlier in the book is that the semantic-like response is algorithmically irreducible to a Turing machine. This assertion is central for our understanding of the laws of the Universe and our comprehension of them. The traditional algorithmic theory is built on the notion of compressibility, i.e. the difference between algorithms has unique quantitative measure so that every algorithm could be reduced to its shortest counterpart. Thus, proclaiming that all laws of the Universe are computable by means of a Turing machine, we proclaim existence of a universal law to which the compressibility reduces each particular one. Yet, our intuition is in strong disagreement with such statement because we "feel" the difference among the laws. Next let us remind the more rigorous arguments that define the impossibility of reducing the laws of Nature to a Turing machine.

One of the most fundamental problems in the information theory is to what extend the information and its processing is related to the organization and functioning of the natural processes. The enormous development of the information theory culminates in the recently widely discussed matter whether the laws of the Universe are computable. The extreme importance of that matter is provoked by the universality of Turing machines which comes to say that every law, viewed as a quantified relation between certain variables, expressed in a closed form, is computable. Then, an immediate outcome of the traditional algorithmic theory is that the variety of laws turns compressible (reducible) to its shortest form so that there is no shorter algorithm executable on any Turing machine. In turn, the existence of universal algorithm (shortest program) whose major property is its insensitivity to the hardware (Turing machine) implies that our Universe is subject to universal law so that each of its constituents obeys it irrespectively to its specific properties. Thus the algorithmic compressibility smears out any difference between one specific property and another.

Though the idea of a universal law has been appealing during the times, it still remains counter intuitive because we "feel" the difference between specificity of different events and phenomena as well as their uniqueness and their irreducibility to one another. Then a closer look on the idea of a universal law displays its inherent controversy. The first one in the line of several problems looks like a technical one: how to compress a stochastic sequence where the response is released from any

pre-determination other than boundedness? As we have already demonstrated in the Chapter 1 there exists a universal ban over any predictability of the nearest and any distant behavior of every complex system. Consequently, the ban over predictability puts a ban on any compressibility in its traditional understanding.

The second controversy commences from the existence of emergent phenomena. Although we are used to associate the emergent phenomena with highly complex systems, actually emergent properties and phenomena occur at every level of matter/energy organization. Thus, an example for emergent property is the glitter of some metals: not a hint of a glitter is embedded in the Hamiltonian through which we can formally describe the properties of any metal. Another example for emergent phenomenon is the emergence of wave-like behavior in Belousov-Zhabotinsky reaction. The self-organized pattern emerges spontaneously from a homogeneous solution. What is important with it is that the spontaneous pattern formation implies anti-thermodynamic behavior, i.e. the emergence goes from a state of maximum entropy (homogeneous solution) to a state of lower entropy (self-organized pattern). Therefore, the spontaneous pattern formation violates the Second Law. Note, however that this violation happens only when external parameters are fine-tuned to certain specific for the solution variables. Thus, once the Second Law holds, the next time it turns violated. Then, how the violation of one and the same law along with its validity would be incorporated in a single Universal law?!

In addition, what is the most counter-intuitive is that an universal law should comprise not only both the violation and the validity of the Second Law, but the glitter of the gold along with the structure and functioning of our mind because it is made of the same atoms and molecules as other objects in the Nature and thus it must be subject to the same law. The latter brings about a vicious circle: It is well known that the "decoding" of any information in the traditional information theory is implemented by an "external mind" which is practically our human mind. Thus, putting it on the same ground as the subjects that it is supposed to decode leads to impasse – who then would be able to decode the information independently from the subjects of decoding. Thus, even though we supposed a universal law to exist, we are not able to recognize it.

Our answer to the problem about the irreducibility of a semantic-like response to a Turing machine is grounded on the substantiation of algorithmic non-reachability of one semantic unit from any other. Our suggestion is that this happens through the presence of a non-recursive component in the discrete band of the power spectra. A great merit of our approach is that the presence of a non-recursive component has been proven as a generic property for the atomic level of hierarchically organized systems. It has been also proven that the appearance of a non recursive component occurs along with the substantiation of an "alphabet" under the mild restraint of spatial heterogeneity. To remind, we found out that on the first hierarchical level, the basins-of-attraction are governed by the control parameters and thus they are firmly set. Then, they could participate to the response by means of associating each intra-basin invariant with a "letter." To remind, the discrete band of a power spectrum appears as intra-basin invariant according to the results obtained in Chapter 4. Therefore, since each discrete band comprises a specific non-recursive component, each "letter" turns algorithmically un-reachable from any other. It is worth noting that at the same time the distance between the "letters" measured in the exerted energy/matter necessary for reaching each of them remains finite and bounded. Thus, the algorithmic unreachability is executed by means of physical reachability.

So far our proof is rigorous only for the atomic hierarchical level where the basic inter-level feedback appears as bounded environment to the self-organization. In turn, the inter-level feedback brings about the non-recursive component in

the power spectrum. The role of the inter-level feedback is highly non-trivial and diverse: (i) it strengthens the response by means of restraining the local fluctuations of concentrations from arbitrary growth; (ii) it is decisive for setting metrics of the state space characterized by inherent for the systems measures; (iii) it substantiates the appearance of non-recursive component in the power spectrum which in turns serves as implement for algorithmic un-reachability. However, our theory can do more: our present task is to prove that it is able to provide algorithmic un-reachability of higher semantic units, e.g., words in a sentence. The matter is highly non-trivial and difficult because the higher level hierarchical structure is highly heterogeneous: it looks like a structured network. The difficulty is increased by the fact the traditional approach to networks brings about rather an appropriate hardware for a Turing machine than to open the door to higher-level semantic-like response.

To remind, Random Boolean Networks (RBN) are known as NK networks or Kauffman networks. An RBN is a system of N binary-state nodes (representing genes) with K inputs to each node representing regulatory mechanisms. The two states (on/off) represent respectively, the status of a gene being active or inactive. The variable K is typically held constant, but it can also be varied across all genes, making it a set of integers instead of a single integer. In the simplest case each gene is assigned, at random, K regulatory inputs from among the N genes, and one of the possible Boolean functions of K inputs. This gives a random sample of the possible ensembles of the NK networks. The state of a network at any point in time is given by the current states of all N genes. Thus the state space of any such network is 2^N. The major property of every RBN is that, on fine-tuning of the input, it reaches an attractor where it stays for arbitrary long time. The remarkable property of that state is its robustness to local failures. In turn the latter makes it very attractive for developing a promising hardware which is insensitive to local failures. This is an example how a network whose elements (nodes) are subject to formal logic reproduces this property at the next hierarchical level. In fact each node could reside in one of two states (active or inactive) and switches between them under the command "IF": "IF" the neighborhood impact reaches certain threshold, the node switches its state. The same is with the attractors: "IF" the local concentration of the nodes in certain state reaches certain value, the system switches to another attractor. Thus, the Boolean networks reproduce formal logic in a way that makes them much more diverse in their abilities and much more reliable in their operations, properties which render them an elaborate hardware for Turing machines.

The importance of Boolean networks for us lies in the following: the invariance of the formal logic as a type of response which appears under self-organization is remarkable! It prompts us to pose the question whether the semantic-like response is also invariant under hierarchical organization. The non-triviality of the matter is set by the fact that along with inter-level feedbacks, a spatially heterogeneous system needs intra-level feedbacks in order to sustain the boundedness of its "nodes." It is obvious that the boundedness of the local fluctuations is necessary for sustaining the stability of a system. Even though they are bounded both in space and in time, their effect is crucial for the stability of organized and structured systems since, being non-linear in general, an interaction between different elements (nodes) could amplify even smallest fluctuations beyond the thresholds of stability. That is why the maintenance of their boundedness is of primary importance for stable functioning of a complex system. Therefore, the task of primary importance is the issue about the intra-level feedbacks organization. Along with this issue comes the question whether or not the stability is exerted by means of semantic-like response of higher order.

Here we have certain repetition of the problem considered in Chapter 3. To remind, we first supposed only boundedness of the velocities of the local fluctuations which has an impact that results in an unrestrained amplification of the local fluctuations. Now we face the similar problem: the Boolean network considers presence of a threshold beyond which the local behavior changes. In turn, the formal logic is reproduced. What else is necessary to produce semantic-like response and is it natural ingredient of the concept of boundedness?

"Mass," "Energy," and "Action" in a Semantic State Space

The setting that a complex system responds in a semantic-like way calls for substantiating semantic units as autonomous objects. Here the autonomy implies existence of properties that characterize the functionality of the corresponding unit as a unique object. To certain extent, this requirement is similar to the notion of mass and energy in the traditional physics where these notions are basic characteristics of each and every object. The notions of mass and energy are central for the traditional statistical mechanics since their participation in the equations of motion is crucial for the understanding of the traditional physics. To remind briefly, the mass explicitly enters the equations of motion while the energy has broader use. In mechanics it is associated with the notion of Lagrangian so that the equations-of-motion to follow from its existence. The relation between the equation of motion and the Lagrangian are given by Euler-Lagrange equations whose origin is set by the Hamilton principle. The Hamilton principle asserts that a trajectory with fixed ends follows the one whose action is stationary.

Even though this result has been considered as one of the greatest results in the traditional mechanics, which strongly impacts all branches of modern physics, we cannot apply it without essential modification to the motion in a semantic state space. There is fundamental reason for this to be the case because, unlike the traditional motion, the motion in a semantic state space is not continuous and thus one cannot fix the ends of a trajectory with arbitrary accuracy. To remind, the trajectories in a semantic state space are bounded irregular sequences (BIS) which are first kind discontinuous curves in every point, i.e. at each point there is a jump of bounded size. Two questions arise: how far the trajectory could go and how close to a certain point it could pass? We have already answered the first of these questions: starting anywhere is the state space the trajectory could reach the thresholds of stability where it exerts a U-turn. Our present task is to establish how close to a certain point it could pass and to find out the measures of this distance. Let us start with the observation that the boundedness sets the following inequalities at each point of the state space and for every trajectory. Let Δx be the "jump" between any two successive states; Δp to be the rate of that jump and Δt to be the time interval at which the jump takes place. Then, since each of these quantities is bounded, we can define the following inequalities:

$$\Delta x_i \Delta x_j \leq C_{xx} \tag{7.1}$$

$$\Delta x_i \Delta p_i \leq C_{xp} \tag{7.2}$$

$$\Delta p_i \Delta p_j \leq C_{pp} \tag{7.4}$$

These relations very much resemble the famous Heisenberg relations in the quantum mechanics. To remind, the Heisenberg relations in the quantum mechanics set a relation between uncertainty of simultaneous measuring of the position and the velocity of a quantum object. This type of uncertainty comes out from the dualism particle-wave, a fundamental notion in quantum mechanics. The general conclusion that stems from the Heisenberg relations is that it is impossible to measure

with finest precision both the position and the momentum of any quantum object.

Then, how we should understand the relations (7.1)-(7.3)? In the present case there is no particle-wave dualism. The answer is simple though non-trivial: we relate these uncertainty relations with the discontinuity of the motion. To compare, the motion in the traditional mechanics is everywhere continuous whereas in a semantic state space it is discontinuous in every point. The other ingredient for the uncertainty relations (7.1)-(7.3) is the boundedness: namely the boundedness sets the values of the constants C_{ij}. It should be stressed that without boundedness, no "uncertainty" relations of the above type are ever available.

However, the relations (71.)-(7.3) comprise more than just setting the uncertainty of determining the position and the velocity in every moment. Indeed, let us define the following variable:

$$\Delta S \equiv \Delta x \Delta p - \Delta E \Delta t \tag{7.5}$$

which we put subject of the following constraint:

$$\Delta S \equiv 0 \tag{7.6}$$

Thereby, we could call S from (7.5)-(7.6) "action." And it is obvious that it retains the property of being stationary (expressed through (7.6)) if and only if the variable ΔE is defined as:

$$\Delta E = \frac{\Delta x \Delta p}{\Delta t} \tag{7.7}$$

Note that ΔE from (7.7) has the meaning of "energy" if defining "mass" as follows:
Defining:

$$V = \frac{\Delta x}{\Delta t} \tag{7.8}$$

---

and

$$\Delta p = mV \tag{7.9}$$

then:

$$\Delta E = mV^2 \tag{7.10}$$

Thus, by means of defining stationarity of the "action," we simultaneously define "mass" and "energy" of the object that follows the trajectory of stationary action. Put it other way round we assert that the motion along a trajectory of a stationary action could be considered as a motion of well defined single object whose constituents move coherently so that to preserve the covariance of the characteristics of the motion. Note, that though dependent on the position in the state space, neither the "mass" nor the "energy" selects a special time or space point along a stationary action trajectory. Thus, the stationary action trajectories still retain the major property of any other trajectory in a semantic state space, namely the property not to select and/or signals out any specific time and/or space point.

Further, it is important to stress on the non-extensivity of the above defined mass and energy. Note, that here neither of them is reducible to sum of mass/energy of the constituents! On the contrary, each of them is explicitly related to non-extensive collective variables characterizing the system. The important point is that defined through stationarity of the action, each of them serves as a characteristic of a motion of a single object. This property serves as grounds for substantiating the idea of a "module" as a functional unit, which, though being itself comprised of sub-units, behaves as a single object. Then, the notions of mass and energy serve as measures of differentiation of the modules one from another.

It should be stressed that the notion of action in the present context does not imply any extreme properties of motion in the state space. To compare, the traditional notion of action relates its stationarity with the extreme properties of the system, such as the principle of least action, minimum energy etc. Thus, the release of the notion of stationarity of the action from extreme properties opens the door to have many trajectories that fulfill the property of stationarity. Thus it is to be expected diversity of response, i.e. diversity of semantics. It should be stressed that, on the contrary, the association of the stationarity of the action with the extreme properties most probably selects a single available trajectory. Thereby, the response of the system, viewed as equations of motion in the state space, is unified and thereby it is not to be expected manifestation of any diversity and non-extensivity.

Before elucidating the role of the presently involved characteristics of the motion with respect to the characteristics of highly organized semantics, let us draw some attention to the relations (71.)-(7.3) and (7.10). Indeed, the boundedness expressed through (7.1)-(7.3) sets boundedness over the velocity V and to the mass m. Then, Equation (7.10) could be presented in the following form:

$$\Delta E = \Delta m C^2 \tag{7.11}$$

where C is the maximum constraint over the velocity along a trajectory of a stationary action. It is obvious that (7.11) surprisingly exactly matches the famous Einstein relation between mass and energy. Our aim now is to focus the attention on the differences with the Einstein formula. The first one is that the bound C is specific to every process (system) and so it does not set any universal constant unlike the case with the Einstein relation which put the light in a special position among all process in Nature. The second fundamental difference between the Einstein relation and (7.11) is that, though (7.11) quantifies transformations between "mass" and "energy," it happens under the constraint of keeping each of them bounded. On the contrary, the Einstein relation allows the mass to grow unrestrictedly when the velocity approaches the speed of light. Indeed, the famous relations $E = mc^2$ where $m = m_0 \big/ \sqrt{1 - v^2 \big/ c^2}$ allows unrestrained growth of both energy and mass when the velocity v approaches the speed of light c. However, since this relation is supposed universal, i.e. available for every system regardless to its particularities and since infinite energy means infinite work for its achievement, does the acceleration process which substantiates this relation implies involving of a special kind of "perpetuum mobile"? This line of suggestions and the relation between (7.11) which, on the contrary, allow transformations of only bounded amounts of energy and mass, and the idea of ban over the perpetuum mobile will be considered in Chapter 9.

EQUATION OF MOTION

A "Word" as a Semantic Unit

Our assertion that the semantic-like response is executed by natural processes is so far confirmed only for the very primitive form of semantics, namely writing "words." Thus, the problem we face now is how words are organized in sentences. For this purpose it is necessary that, when a "word" is "written," i.e. the corresponding trajectory ends at "space bar," the next word not to be "picked" randomly. It is obvious that a special choice of the next word would be provided if the motion in the state space comprises a "flow" component. Then, since the flow is directed, it will impose long-range correlations among distant words. Further, since the sate space is bounded, any trajectory that "writes a sentence" will come back to the initial point, which will thus play the role of a "dot." Thereby, the accumulation points can serve both as space bars and dots.

But how a flow will come? This is the question we will consider next. We have already established that the "letters" in an "alphabet" are associated by the intra-basin invariants of the self-organization at the most basic hierarchical level. We have demonstrated that the intra-basin invariants are presented by the discrete bands of the power spectra of the corresponding solutions of the equations that govern the evolution of the self-organization at the basic hierarchical level. A great achievement of our approach is that the solution of these equations (Equation (4.1)) retains the generic property that its solution is BIS whose power spectrum is additively decomposed into a specific discrete band and universal continuous one. An exclusive property of the power spectrum is that along with these bands it comprises a non-recursive component which is brought about by the interplay between the boundedness of the periodic motion featured by the discrete band and the boundedness of the stochastic motion presented by the continuous band. The presence of a non-recursive component is crucial for the discrimination of any semantic-like response from a Turing machine. In the Background to the present Chapter we have already considered the role of non-recursive component for this discrimination. Thus, we face the question whether the higher order semantics, e.g. writing sentences retains this property, i.e. whether the higher order semantics is also irreducible to a Turing machine?

Let us now start with the observation that every word is "written" by letters. Thus, a "sentence" is a trajectory in the state space which is subject to special long-range relations. Hence, the following question arises: is this long-range order of "letters" subject to any general constraint? Next we will demonstrate that such constraint do exists and it is the constraint of the stationarity of action. In the previous section we have considered the properties of the action in a semantic state space. We have found out that its exclusive property is that it is not associated with any extreme properties of the motion but is associated with specific "mass-energy" transformations quantified by the relation (7.11). A crucial for the further considerations property of this relation is that it sets properties of certain "flow" in the state space. Note that each of the variables "mass" and "energy" comprises characteristics of the motion (see (7.8)-(7.10)) and their transformations from one into another are subject to specific quantification. Thus, the stationaity of the action is indispensably related to the existence of a flow in the state space. This opens the question whether there is a general equation which governs the motion on those trajectories where the action is stationary and if so what is the role of the flow?

The starting point for answering this question is the fact that every trajectory in the state space is a BIS, subject to the additional constraint of the stationarity of its action. In order to highlight better the role of the flow let us start the quest for a general equation of motion with the simple constraint of boundedness of the successive "jumps." If the motion is subject to this constraint only, the following equation of motion is available:

$$\frac{dx}{dt} = \bar{V} + \mu(x) \qquad (7.12)$$

where $\bar{V}$ is the expectation value of the velocity over the trajectory and $\mu(x)$ is a zero-mean BIS which takes into account that every sample average deviates from the expectation value by a finite, yet bounded value $\mu(x)$. Therefore, obviously no higher order semantics is to be expected.

Let us now consider presence of a flow. Then the balance equations cast as follows:

$$\frac{dx}{dt} = V(x) + \mu(x) + \nabla \bullet \left(D(x) \nabla x \right) \qquad (7.13)$$

where $D(x)$ is the diffusion coefficient; $V(x)$ is the average over a basin-of-attraction velocity; $\mu(x)$ is the deviation of the current velocity from

$V(x)$. The reason for replacing the expectation value of the velocity with the average over a basin-of-attraction comes from involvement of the flow. Indeed, its presence makes the equation of motion (7.13) non-linear partial differential equations and thus it is to be expected that their solutions are specific for certain domain of control parameters. In order to set the corresponding domains, it is necessary to provide a procedure of constructing (7.13) which sets the values of the corresponding terms in a non-ambiguous manner. The necessity of such procedure is brought about by the extreme sensitivity of the position of the domains to the values of the terms participating in any non-linear equation. The equation of state (4.1) is grounded on the unique choice of units as discussed in section *Metrics in the State Space* in Chapter 4. Accordingly, the metrics is preserved under the use of the following procedure of constructing the equation (7.13): given that the basic trajectory passes through several basins-of-attraction. It is obvious that every averaged over a basin-of-attraction velocity turns out an intra-basin invariant. Thus, since the velocity that participates (7.13) is the current one, it could be presented as a sum of averaged over the basin of attraction one $V(x)$ and the deviation from it $\mu(x)$. To compare, the velocity $\bar{V}$ which participate (7.12) is the expectation value, i.e. it is averaged on arbitrarily large time interval which obviously comprises all basins-of-attractions. Further, it is worth noting that if we use $V(x)$ as the average in (7.12), it will yield reproduction of the basic equations (4.1) and thus one cannot expect any higher organization of semantics.

It is obvious that Equation (7.13) is a bounded stochastic partial differential equation. It is of the same type as equations (4.1) and thus it is to be expected that it retains similar properties.

The most important conclusion of the similarity between Equation (7.13) and Equation (4.1) is that it reproduces the major properties of a semantic unit at a higher level of organization. According to it, each "word" is expressed through a spatio-temporal pattern which characterizes a given basin-of-attraction. And again, only transitions between adjacent basins-of-attraction are admissible. In turn, not every word could follow each other. Thus, like the letters in a word, the words in a sentence follow certain rule.

However, the most important property of the similarity between Equation (4.1) and (7.13) is that now we have a pattern which opens the door to further hierachization of the semantics. By means of defining stationary action according to (7.13), we could produce the next level equations of motion which would be of the type (7.13) and thus its reproduction produces further hierachization of the semantics.

The major question arises: how to construct equations-of-motion at each hierarchical level? The answer to this question comes from our result that we can define the notion of 'energy' so that we can associate a Hamiltonian with that 'energy'. In turn, the corresponding Hamiltonian is the major implement for substantiating specific rules which local interactions obey. According to the results proved by Turing (Turing, 1952), a necessary condition for keeping the result of interactions locally stable and independent from the "history" of the individual interactions is sustaining the unitarity of the local dynamics. This is achieved if and only if local Hamiltonians are Hermitian. Then, the unitary evolution substantiated by means of Hermitian Hamiltonians provides linear superposition of the interactions, i.e. the result of a complex interaction is linearly decomposable to its ingredients and stays stable if the latter are stable. However, if not bounded, the linear superposition cannot prevent uncontrollable growth of local accumulation of matter/energy. The way out from this situation comes when considering the role of the non-unitary interactions as it has been made in Chapter 3.

The conclusion drawn from the above considerations is that the abstract notion of a Hamiltonian and unitary evolution are available at every hierarchical level. Their role is to set specific rules for local interactions so that the result of the latter to stay stable in the sense of Turing. Moreover, the unitarity of the evolution and the obtained stability renders assigning a time-independent probability to the outcome of each event described by the corresponding local Hamiltonian. In turn, the setting of time-independent probabilities to the local event allows the use of all developed so far apparatus of the traditional statistical mechanics for constructing the equation-of-state at a given hierarchical level. Yet, it should be stressed that this apparatus is available for derivation of the corresponding equations at the hierarchical level itself. The inter-level feedbacks and the non-unitary interactions should be taken into account by means of strategy developed in Chapter 3.

It is important to stress that the reproduction of the equation (7.13) on higher levels justifies the ubiquity of the property that every semantic unit is characterized by a spatio-temporal pattern whose homeostasis is described by a specific discrete band in its power spectrum. Alongside, a prominent property of the hierarchical organization is that the one set by it homeostasis is characterized by the presence of a non-recursive component along with the discrete and the continuous band of the shape $1/f^{\alpha(f)}$. This gives a definite answer to the question whether a semantic-like response is reducible to Turing machine and the answer is definitely negative: no semantic-like response is reducible to a Turing machine!

Another general outcome of the similarity between Equation (7.13) and (4.1) is that about the non-extensivity of the semantics. Although similar, these equations are not reducible from one into another! It is worth noting that Equation (7.13) is subject to the additional constraint of the stationarity of the corresponding action; and each higher level semantic organization is subject to its own constraint of a stationary action. Therefore, each word, viewed as a semantic unit of a sentence, has different meaning and different role than the same word viewed as a sequence of letters.

It should be highlighted the difference between the semantic-like response and Turing machine: the first fundamental difference is that the semantic-like response is self-organized and autonomous whereas Turing machine is non-autonomous, i.e. it "writes" and "decodes" the "message" only by means involving external mind; the second fundamental difference is that the semantic-like response is non-extensive whereas the messages produced by a Turing machine are always extensive. Note, that the message could be considered non-extensive only by means of strong interference of an external mind and always in the context of a certain frame created also by that external mind; last but not least, while the computability by Turing machine implies algorithmic reachability of each law from any other, the ubiquitous presence of a non-recursive component in the characteristics of each semantic unit implies uniqueness and algorithmic unreachability of each law from every other. It is worth noting that the above considerations render Turing machine and semantic-like response rather counterparts than opponents. Indeed, a Turing machine can accomplish every task defined in a recursive way whilst a semantic response cannot. At the same time whilst the semantic response has autonomous comprehending, the output of a recursive computing needs "external mind" for its reading.

CONTROL OVER NETWORKS

Matter Wave Emitting

An obvious requirement for launching a semantic response is that the corresponding "alphabet" comprises no less than 3 "letters." It appears as necessary and sufficient condition for preventing formation of "words" which comprise arbitrary combinations of "letters." The sufficiency is

brought about by the fact that the motion in the state space goes through adjacent basins-of-attraction only. Therefore, the semantic response is substantiated only by words whose sequence of letters meets this condition. The necessity becomes obvious by the following consideration: an 'alphabet" of two "letters" cannot automatically discriminate between a random sequence and a semantic one; on the contrary, an alphabet of 3 "letters" allows such discriminations because each and every allowed succession of letters goes only through adjacent basins-of-attraction; thus there is a natural ban over random sequences. The necessity of having at least 3 "letters" in the "alphabet" imposes the condition that the corresponding equation-of-state to have at least 3 basins-of-attraction. Thus, we face the question whether there is a general type of equations which provide this condition fulfilled. The answer is positive: the non-linear partial differential equations are the desired type of equations that can serve as equations-of-state providing at least 3 basins-of-attraction. It should be stressed that the non-linear ordinary differential equations are not able to fulfill the task because they have only two types of solutions: fixed point and self-sustained oscillations. Thus, they could serve as hardware for a 2-bit Turing machine. Yet, as we have already highlighted, a 2-bit computing is not a semantic one because it is impossible to define a "space bar" as a separator of words. Indeed, in a two- "letter" space the "space bar," viewed as the point tangent to both letters, separates each letter from its predecessor and successor. Thus, the set of 'words" is reduced to a set of two "letters." Thereby, it is obvious that it is impossible to define any semantics out of 2 "letters" only.

The necessity of using partial differential equations, however, implies presence of spatial heterogeneity. The question is whether this spatial heterogeneity is local, i.e. it is to be associated with the functioning of a node or it should be considered as a condition for specific spatial extension of the "network." This question is posed by the following consideration: choosing the first alternative, i.e. assuming that the spatial heterogeneity applies to a node only, we come to the idea of a Boolean network: the state of a node switches from "active" to "inactive" depending on the total environmental impact. Thus, the "informational state" (active, or inactive) can be effectively described by a system of ordinary differential equations. Thereby, regardless to the physical nature and the structure of the nodes, the information state space is always a 2-bit one.

These considerations pose the question whether there is another way of taking into account the spatial heterogeneity. Our next task is to demonstrate that the spatial heterogeneity brings about a mechanism that controls the homeostasis in a network which in turn justifies the use of bounded stochastic partial differential equations as equations-of-state. Let us start with considering the effect of the environmental impact on a node: the traditional perturbation approach considers it as a linear superposition. However, it is available only if there is no threshold which separates two states (active and inactive). Indeed, a closer look shows that the effect of impact depends not only on its total sum but on the succession of application of its terms. Thus, for example, even if the sum does not indicate "switching," it could be a result of the presence of two terms of opposing signs such that each of them is large enough to produce a "switch" alone. This circumstance provokes high instability of the network and high sensitivity of its functioning to any "noise." Then, the network would be particularly vulnerable by the unavoidable exposure the fluctuations produced at the most basic hierarchical level. Then, in order to suppress this vulnerability, the network must have mechanism that sustains its robustness to small environmental impacts, i.e. to sustain its local homeostasis.

Next we will demonstrate that the most general form of the partial differential equations has a type solution which has not been considered so far. The most general form of partial differential equations is as follows:

$$\frac{\partial \vec{X}}{\partial t} = \vec{\alpha} \bullet \hat{A}\left(\vec{X}\right) - \vec{\beta} \bullet \hat{R}\left(\vec{X}\right) - \nabla \bullet \left(\hat{D}\left(\vec{X}\right) \bullet \nabla \vec{X}\right)$$

$$(7.14)$$

where $\vec{\alpha}$ and $\vec{\beta}$ are the control parameters; $\hat{D}\left(\vec{X}\right)$ is the matrix of the diffusion coefficients; and $\hat{R}\left(\vec{X}\right)$ are the stock and sinks functions.

In order to sustain permanent boundedness over the local accumulation of matter, the solution of Equation (7.14) must be cast in the form:

$$\vec{X}\left(\vec{r}, t\right) = \int \int \left(X_{\omega,k}^{\mathrm{Re}} + i X_{\omega,k}^{\mathrm{Im}}\right) \exp\left(i\omega t + i\vec{k} \bullet \vec{r}\right)$$

$$(7.15)$$

where the imaginary part $X_{\omega,k}^{\mathrm{Im}}$ is the part of the solution $\vec{X}\left(\vec{r}, t\right)$ supposed to provide emitting of the matter away from the given locality. The comparison between the real and imaginary parts of the l.h.s. and r.h.s. of Equation (1) sets that the conditions for $X_{\omega,k}^{\mathrm{Im}}$ to be always non-zero are:

$$\omega + \vec{k} \bullet \nabla \hat{D}\left(\vec{X}\right) = 0$$
$$i\vec{\alpha} + i\vec{k} \bullet \hat{D}\left(\vec{X}\right) \bullet \vec{k} = 0$$

$$(7.16)$$

In order to make the calculations explicit, Equation (7.16) is carried out for $\hat{A}\left(\vec{X}\right) = \hat{1}$ and zero linear part of $\hat{R}\left(\vec{X}\right)$. It is obvious, that the Equation (7.16)) select unique non-zero values for ω and $\vec{k}$ which defines a wave that spreads with finite velocity. It should be stressed that the necessary condition for setting a well defined finite velocity of the wave is the boundedness of the local gradients (expressed through the gradient of the diffusion coefficient matrix $\nabla \hat{D}$). Note, that the fine-tuning of the matter wave to the control parameters is implemented through the dependence of the wave-vector on the control parameters $\vec{\alpha}$.

Outlining, the spontaneous emitting of a specific matter wave appears as a generic property of the most general form of the partial differential equations.

It is worth noting that the property of emitting a matter wave serves as a control mechanism for sustaining homeostasis in the presence of noise as well. Indeed, recalling that the presence of noise is formally correspondent to a specific shift of the control parameters in Equation (7.14), it is obvious that thus emitted matter wave will "disperse" accordingly the accumulated by the noise "extra" matter/energy.

Thus one can definitely conclude that the presented by us mechanism for controlling the local homeostasis through matter wave emitting makes the corresponding network stable and robust to small local and global environmental fluctuations. Alongside, it makes possible to define the stochastic bounded partial differential equations as equations-of-state. In turn, the latter opens the door to substantiating an "alphabet of more than 3 "letters" which serve as ground for semantic-like response.

Thus we can write down explicitly the most general form of the equation-of-state of a hierarchically organized system:

$$\frac{\partial \vec{X}}{\partial t} = \vec{\alpha} \bullet \hat{A}\left(\vec{X}\right) - \vec{\beta} \bullet \hat{R}\left(\vec{X}\right) - \nabla \bullet \left(\hat{D}\left(\vec{X}\right) \bullet \nabla \vec{X}\right) + \mu\left(\vec{X}\right)$$

(7.17)

where $\mu\left(\vec{X}\right)$ is the "noise" term.

It should be stressed on the difference between the considered above matter waves and the waves produced by self-organization: while the above waves are fine-tuned to the current control parameters, the characteristics of the self-organized waves are intra-basin invariants. This difference and the purpose of the emitted mater waves to sustain the local homeostasis allow us to consider them as intra-basin feedbacks among distant nodes. Let us now make a parallel between their role and the role of the non-unitary interactions introduced on the basic hierarchical level. To remind, the non-unitary interactions appear as a result of a non-trivial interplay between the random and the potential motion of atoms and molecules. There major property turns to be the broken time-symmetry of interactions which makes the corresponding Hamiltonian dissipative. In turn, the feedback between it and the local gapless modes, e.g. acoustic phonons, renders "dispersion" of the locally accumulated extra matter/energy. The consideration of this feedback in the frame of boundedness renders the covariance of the obtained bounded fluctuations, i.e. the characteristics of neither fluctuation depend on its location in the space and time. Similarly, an emitted matter wave, viewed as a general solution of the equation-of-state also does not involve explicitly specific time and space point for its emitting which could be considered as retaining of the property of covariance. It should be stressed on possible confusing of the matter about covariance of matter wave emitting and the specific parameters of its emitting in every concrete case. Note, that the covariance implies lack of any explicit involving of a space/time point. The fact that this condition is fulfilled in every concrete case comes out from the following consideration: the direction of emitting is controlled by the current value of the control parameters and the current gradient of the diffusion coefficient and thus the parameters of the emitted wave do not involve *explicitly* any specific time/space point, although each matter wave is emitted in a specific time and in a specific direction.

It is worth noting that the obtained above covariance has much deeper origin because it is rather manifestation of the time-translational invariance of the evolutionary dynamics. Indeed, the "bounded randomness" of the rates of mass/energy exchange and the boundedness of the local instabilities serves as implements for establishing local conservation laws in the form of "continuous field" equations for the dynamics of the relevant variables (Equation (7.17). Further, as proved above, the mechanism of sustaining local homeostasis by means of matter wave emitting maintains the description of the local and global evolution in terms of continuous field equations. Then, according to the Noether theorem, the evolutionary dynamics of the relevant variables is time-translational invariant. The novel point is that now the "field" form of the relevant variables arises *spontaneously* from the dynamics modified accordingly within the frame of the boundedness concept. To compare, the traditional statistical mechanics *postulates* the "continuous field" form of the relevant variables along with suggesting decoupling of the dynamics into random motion and Hamiltonian dynamics. As proved in Chapter 3, the destructive flaw of this setting is that the suggested decoupling allows unlimited amplification of the local instabilities which not only endangers the stability but also discards any "continuous field" form of the relevant variables. Alongside, as proven above, the consideration of the environmental impact by means of linear superposition yields high instability of any network.

A SEMANTIC UNIT AS A SPATIO-TEMPORAL PATTERN

So far we have established that the equation of state for each node in a network is given by the most general form of bounded stochastic partial differential equations (7.17). Since they are non-linear differential equations it is to be expected that their solutions are different in different basins-of-attraction. Since they comprise a spatially extended term (diffusion), the corresponding solutions cast in the form of specific spatio-temporal patterns. A crucial question arises: what is the role of the "noise" term in Equation (7.17); is it similar to the role it plays in Equation (4.1)? At first, let us remind that a spatio-temporal pattern is characterized not only by its power spectrum but also by its specific structure whose measure

is the so called form-factor. The latter is a characteristic that measures the structure of a pattern through measuring the relative distance among its constituents in an experiment on bombarding the pattern by rays of different nature, e.g. X-rays, neutron scattering etc. Next we will demonstrate that the noise term gives rise to two effects: (i) the form-factor of every scattering is additively decomposed to a specific discrete and a universal continuous band of shape $1/k^{\alpha(k)}$ where k is the wave vector; (ii) the discrete band comprises a non-recursive component. The proof is explicitly grounded on the fact that every spatio-temporal pattern is comprised by components each of which is a BIS. It should be stressed that the boundedness of the fluctuations in a pattern is maintained by means of two general mechanisms: (i) the inter-level feedbacks which keeps the fluctuations of the concentrations bounded; (ii) intra level feedbacks which maintain the effect of inter-node interactions bounded by means of emitting a matter wave. We pose the question whether the inter- and intra-level feedbacks are enough to provide separating of the genuine information encapsulated in the discrete band from the "noise" which comes from the hierarchical structure. Further, the presence of a non-recursive component is decisive for the general view of the present book about the algorithmic unreachability of one information unit from another. We start our proof by considering the first part of the question, that of the additivity between a discrete and a continuous band in the form-factor.

A key problem of processing information is outlining the general conditions that provide the separation of the genuine information from the accompanied noise. Here this problem arises from our assumption that a steady state in an ever-changing environment is represented by bounded irregular sequences (BIS) among which are sequences that comprise both discrete bands (information) and continuous bands (noise). It is obvious that the reproducibility of the information is met only if the accuracy of extracting information is independent from the statistics of the noise. The ontological aspect of the problem is that the target independence is an indispensable part of preventing the interference between the causal relations that bring about information and the noise. Indeed, note that while the information remains the same on repeating the same external stimuli, the noise realization is ever different. The task of this section is to demonstrate that the problem is successfully resolved in the frame of the proposed by us concept of boundedness.

Let us remind that the boundedness of Equation (7.17) implies that each component of every solution of the system is a BIS which comprises both a discrete and a noise band in its power spectrum. Note that when corresponding BIS concerns spatio-temporal pattern its power spectrum is replaced by the notion of form-factor. We will prove that the discrete and the noise band of every such form-factor are additively superimposed; moreover the shape of the noise band is insensitive to the statistics of the "noise" (environmental variations) and to the status of the recipient. More precisely, it could be only the shape $1/k^{\alpha(k)}$ which exactly matches the properties of $1/f^{\alpha(f)}$ noise established before.

Another important aspect, related to the idea of information viewed as a measure of causality, is that the additivity of the noise and the discrete band along with the insensitivity of the shape of the noise band to the statistics of the noise are the necessary and sufficient conditions for preventing the interference between causality and noise. So, indeed the boundedness is the target general condition that provides reproducible separation of the genuine information from the noise. But now we shall prove more: we shall prove that the only decomposition of the power spectrum compatible with the boundedness is the additivity of the discrete and continuous band. To prove that assertion, suppose the opposite, namely: the

discrete and the continuous band are multiplicatively superimposed:

$$A(x) \propto \int_{k_0}^{\infty} g(k) \exp(ikx) dk \sum_{l=1}^{\infty} c_l \exp(ilx)$$

(7.18)

where $A(x)$ is the amplitude of the variations in the time series; $g(k)$ are the Fourier components of the continuous band and c_l are the Fourier components of the discrete band. The purpose for representing the multiplicative superposition of the discrete and continuous band through their Fourier transforms is that it makes apparent the permanent violation of the boundedness of $A(x)$ even when both the noise and the information are bounded. Indeed, it is obvious that there is at least one resonance in every moment regardless to the particularities of the causal relations between c_l and the noise $g(k)$. In turn, the resonances break the boundedness of $A(x)$ because they make the local amplitude to tend to infinity. This, however, violates energy conservation law because it implies that finite amount of energy/substance (bounded information and bounded noise) generates concentration of infinite amount of energy/substance (infinitely large amplitude of fluctuations at resonance).

However, it is worth noting that alone the additive superposition of a discrete and a continuous band is not enough to provide the reproducibility of the information. Indeed, the additive superposition reads:

$$A(x) \propto \int_{k_0}^{\infty} g(k) \exp(ikx) dk + \sum_{l=1}^{\infty} c_l \exp(ilx)$$

(7.19)

It is obvious that the reproducibility of the genuine information encapsulated in c_l requires insensitivity of $g\left(k\right)$ to any noise realization, i.e. to the noise statistics. As we have already proved in Chapter 1, this is possible only for bounded noise; then the continuous band in its power spectrum fits the same shape regardless to the particularities of the noise statistics; and as we have already proven this shape is $1\Big/k^{\alpha(k)}$.

Summarizing, we can conclude that the boundedness is that single general constraint which provides preventing interference between causality and noise in a self-consistent way. The key importance of that result is its immense ubiquity. Since the boundedness as the constraint that prevents interference between causality and noise is independent from any particularities of the processes that generate time series, it is equally available for systems of different nature ranging from cosmic rays to DNA sequences and financial time series to mention a few. Alongside, additive decomposition of the form-factor to a specific discrete and a universal continuous band is insensitive to the nature of the incident radiation, i.e. it does not depend whether we measure the structure by means of X-rays, neutron scattering etc.

Now we are ready to answer the question whether the interplay between the boundedness of the discrete part of the solution and the boundedness of the continuous one brings about a non-recursive component. Next we will prove that the same origin of both discrete and continuous bands in the form-factor, i.e. the fact that each of them comes from the same equations (7.17) is justified by the presence of a non-recursive line arising from their interplay. The interplay commences from the non-linear and non-homogeneous way of participation of the inter-level and the intra-level feedbacks in (7.17). And since the overall solution of (7.17) is bounded irregular function it could be presented as wild and permanent bounded "twisting" and "winding up" around a "skeleton" built on the discrete band of the power spectrum and the form-factor. Thus, the motion resembles the

motion on a torus with irrational frequencies. In the present case the motion never stops and thus extra-line associated with irregular motion around the skeleton appears always as non-recursive.

It is important to stress that the non-recursive component must not be viewed as a component that gives certain distance between constituents of any given system. It is obvious that any such consideration would yield an absurd, i.e. points with arbitrary large concentration of matter alters points of zero population, a situation which is in apparent contradiction with the boundedness. Instead, an alternative explanation given by our approach is that the non-recursive component is to be considered as a measure for the threshold over the local deviations from the pattern "prescribed" by the discrete band. Alongside, being an exclusive property of the solution of Equation (7.17) the presence of a non-recursive component in the power spectrum and in the form-factor serves as a hallmark of presence of hierarchical order in a studied system.

Yet, the greatest advantage of the above general properties of every spatio-temporal pattern is that it could serve as a semantic unit. Each spatio-temporal pattern provides the following properties to the corresponding semantic unit: (i) defined through the power spectrum and/or form factor it appears as an inter-basin-invariant; (ii) the ubiquitous presence of a non-recursive component provides the algorithmic un-reachability of one semantic unit from another; (iii) along with the algorithmic unreachability, the physical reachability of adjacent semantic units is provided.

It is worth noting that along with the above properties, the hierarchical organization provides non-extensivity of the higher order semantic units. Since the semantic trajectories on each hierarchical level are selected under the constraint about the stationarity of the action, not a single semantic trajectory is algorithmically reducible to a trajectory at the lower hierarchical levels.

Scale-Free Property of Networks

So far we have established that the most distinctive property of a hierarchically self-organized network is that the form-factor of its nodes retains a general property of comprising a non-recursive component. We already have mentioned that this property should be understood as the boundedness over the local deviations of the structural constituents from the specific structure sets by the discrete band in the form-factor. Thus we face the question whether the property to be a scale-free applies to the thereby constructed network. To remind, a scale-free network is a network whose degree distribution follows a power law, at least asymptotically. That is, the fraction $P(k)$ of nodes in the network having k connections to other nodes goes for large values of k as

$$P(k) \cong ck^{-\gamma} \tag{7.20}$$

where c s a normalization constant and γ is a parameter whose value is typically in the range $2 < \gamma < 3$, although occasionally it may lie outside these bounds.

Scale-free networks are noteworthy because many empirically observed networks appear to be scale-free, including World Wide Web links, citation networks, biological networks, airline networks and some social networks.

Like the case of Zipf's law we again encounter the puzzling ubiquitous encountering between the specific structure and functioning of a network on the one hand, and their ignorance by the power law distribution on the other hand. To remind, the Zipf's law establishes power law distribution for the frequency of appearance of a word depending on its rank in the frequency table. Thus, the Zipf's law seems to sweep out any difference between a text and a random sequence of words. In Chapter 6 we have established that the commence of the Zipf's lies in the fact that the association

of a word with the frequency of its appearance implies violation of the long-range correlations among words set by the higher order hierarchical organization of the semantics. Likewise, the association of the structure of a network with its degree distribution has the same effect: a power law distribution appears for the degree distribution at the expense of ignoring the long-range functional correlations between distant nodes. Yet, as considered in Chapter 5, at the same time an empirical observation of scale-free property, viewed as a power law distribution, serves as a criterion for hierarchical structuring. Thus, we can utilize the power of the probabilistic approach for development of that criterion.

Outlining, the scale-free property is another example of the power law distribution that appears as a result of ignoring the long-range correlations among system's constituents. That is why we introduce the property of presence of a non-recursive component as a decisive test for the highly non-trivial role of hierarchical order and the inter-level and intra-level feedbacks.

Through Diversity to Sustainability

We have already established that a semantic-like response is a generic property of systems subject to morphogenesis. We also have found out that the one of the major properties of that morphogenesis is that each and every homeostatic pattern is robust to small environmental changes. However, the morphogenesis itself gives rise to a specific type of vulnerability: it is that the impact of higher hierarchical levels onto the lower ones is locally different. Then, it poses the question that if the local impact is different how the corresponding nodes self-organize themselves so that to provide robustness to local failures at the presence of local diversity of the environmental impact?

In order to answer this question we put forward the idea that the robustness to local failures is provided by means of coexistence of different "mutations" of the same 'kind': each "mutation" respond to specific local environment. Further, we assume that different 'mutations' are linked with one another by means of interactions described by equations of type (7.13). The major question now is to prove that it is possible to derive the notion of a 'kind' that stands behind specific mutations. To prove it we assume that the dynamics of a given 'kind', viewed as the corresponding dynamical system, is equivalent to a reduced system of differential equations according to the Central Manifold Theorem. Thus, the "physical" solution (different mutations) has its mathematical match, namely the corresponding central manifold (corresponding 'kind'). The non-triviality of the present setting is that the hierarchical subordination fixes the uniqueness of the central manifold. In turn the uniqueness of the central manifold provides the robustness of a 'kind' viewed as a complicated functional and structural organization of different mutations, phenotypes, alleles and individuals.

Now we come to the issue about how the idea of mutation provides robustness to local failures. It is well known that the latter is a property of networks. But how the notion of a network is related to the idea of morphogenesis? We consider the relation between the notion of a network and the notion of specific morphogenesis in the following: the specificity of any given morphology lies in its tiniest inter- and intra-level feedbacks which operate among nodes which represent different mutations. Assuming that network describes a 'kind', all nodes of different mutations appear the same thus making the corresponding network to appear as a homogeneous structure. This is exactly the same as with the Zipf's law where the assigning of a probability to a word occurrence ignores correlations involved by the higher level semantics. Here we have the same: the robustness to local failures allows us to ignore the correlations involved by the fine-scale inter and intra-level feedbacks by means of "smearing out" the differences between mutations.

Our idea of associating the notion of a network with the notion of a 'kind' finds great support by the ubiquity and universality of the scaling laws that describe different networks. Next we focus the attention on the question why the biological scaling laws persistently retains exponent that is less than 1 whilst the scaling laws typical for human society persistently retains exponent larger than 1. Our answer to this question is that this difference appears because of the different goals of the biological and cultural evolution. Thus, the biological evolution is focused on preserving homeostasis through minimizing its response to environmental variations while the cultural one is focused on developing non-extensivity of the hierarchical semantic-like organization. The following arguments come in support to this conjecture: the scaling laws can be considered as impact-response relations; then, the smaller than 1 exponent justifies that the response of a more massive object is relatively tightened compared to the response of a less massive object. Put in other words, the focus is on tightening of the homeostasis. At the same time the greater than 1 exponent implies intensification of the non-extensivity of hierarchical super-structuring through implementation of multi-valued use of the same networks and their mutual interactions.

Thus we conclude that the general way of 'building' a hierarchy is through morphogenesis considered in the setting of the developed by us concept of boundedness. Its major advantage is that it provides a sustainable evolution of a 'kind' by means of diversifying its mutations.

CONCLUSION

The major task of the present Chapter has been to demonstrate that the interplay between the boundedness and the hierarchical order is the major implement for reaching diversity of patterns and behavior. The exclusive properties of this diversity are: (i) stability of the behavior of each pattern. The behavior of each pattern is constrained to vary within specific margins dictated by both inter-level and intra-level feedbacks so that not to select any specific time or spatial point; (ii) the diversity of the patterns and their stability is expressed in an intelligent-like way though semantic-like response. Each pattern is to be associated with a semantic unit. An important property of the semantic units is that they are algorithmically unreachable from one another. The algorithmic un-reachability is implemented by the ubiquitous presence of a non-recursive component in each power spectrum and in each form-factor. It is worth noting that the presence of a non-recursive component does not quantify any metric relations among distances but serves as a hallmark for the permanent presence of bounded noise which commences from the inter-level feedbacks.

The exclusive property of the higher order semantic organization is its non-extensivity. The latter is implemented by means of imposing the constraint of the stationarity of the action along the trajectories in the state space. It is worth noting that, unlike the traditional mechanics, our approach does not associate the stationarity of the action with any extreme properties of the system such as minimum energy, maximum entropy etc. In turn, the action acquires the property to be stationary on a number of trajectories. Therefore, it substantiates the corresponding "grammar" by means of imposing specific "directed flow" on "picking up" successive semantic units. Alongside, the release from any association with extreme properties substantiates the diversity of the semantics; therefore there are a lot of possibilities for a semantic-like response.

There is a very important interplay between the diversity of the response, its stability and the non-extensive hierarchical organization of the semantic-like response: though the diversity intensifies with the increase of the hierarchical organization, the behavior of the system becomes more and more predictable. This, to a certain extent surprising result, is evoked by the non-extensivity of the hierarchical organization of the semantic-like response. To make this clear, let us remind that the higher order grammar rules "direct" ordering of the lower level semantic units, e.g. grammatical rules arrange the word order in a sentence. Thus, the knowledge of higher level grammar provides an "expanding" of the range of predictability of the corresponding "Oracle." To remind, we call the property of "foreseeing" the next semantic unit on the grounds of the knowledge about the current one only an "Oracle." Now the "Oracle" could predict/guess not only next/missing letters, but it can predict/guess next/missing word, sentence, et cetera.

REFERENCES

Turing, A. (1952). The chemical basis of morphogenesis. *Philosophical Transactions of the Royal Society B, 237*, 37–72. doi:10.1098/rstb.1952.0012

Chapter 8
Sustainable Evolution in an Ever-Changing Environment

ABSTRACT

It is suggested that the notion of equation-of-state serves as appropriate common basis for studying the macroscopic behavior of both traditional physical systems and complex systems. The reason is that while the equilibrium systems are characterized both by their energy function and the corresponding equation-of-state, the steady states of out-of-equilibrium systems are defined only by their dynamics, i.e. by their equations-of-state. It is demonstrated that there exists a common measure which generalizes the notion of Gibbs measure so that it acquires two-fold meaning: it appears both as local thermodynamical potential and as probability for robustness to environmental fluctuations. It is proven that the obtained Gibbs measure has very different meaning and role than its traditional counterpart. The first one is that it is derived without prerequisite requirement for simultaneous achieving of any extreme property of the system such as maximization of the entropy.

INTRODUCTION

So far we have studied the grounding principles that underlie the functioning and organization of complex systems so that their evolution to be self-sustained in a stable way in a non-predetermined environment subject to the constraint of bounded-ness alone. This setting is fundamentally different from the traditional one which always pre-determines environment thus aiming to provide the conditions for reoccurrence of the event. Thereby, the predetermination of the environment is viewed as necessary prerequisite of a law: a law will state that when an event reoccurs the environment re-

DOI: 10.4018/978-1-4666-2202-9.ch008

occurs too. The present book aims at developing a systematic alternative to this view. That is why it does not involve any pre-determination of the environment as a condition under which the major properties of the complex systems have to appear. The question is how general our approach is – is it general enough to include traditional physical systems as part of a more general scheme and if so what is this scheme?

We have found out that the response of a complex system comprises two general components: a specific one, called homeostatic, which is robust to small environmental variations, and a universal one, whose properties are insensitive to the statistics of environmental fluctuations. It should be highlighted that both components commence from the collective behavior of a system: the homeostatic part of the response is to be associated with the corresponding spatio-temporal pattern whereas the universal part is substantiated by the basic inter-level feedback which spontaneously emerges at every higher hierarchical level as we have demonstrated in the previous Chapter. Alongside, the central for our approach proof that consists of additive superimposing in the power spectrum of the corresponding characteristics of the homeostatic and the universal part of the response renders the constant accuracy of reoccurrence of the homeostasis in an ever-changing environment. Further, by means of delineating the corresponding basin-of-attraction, we are able to figure out the domain of "robustness" of each spatio-temporal pattern which characterizes a given homeostatic pattern. Thus, we substantiate a fundamental generalization of the notion of a law: a pre-determined environment is not necessary for an event to be sustained and to reoccur!

The knowledge about the structure and functioning of the complex systems in a non-predetermined environment poses the question what kind of measure to apply to cases when a system approaches the thresholds of stability by means of exerting fluctuations. At first sight it seems that this task is similar to the traditional task

of finding out the probability for a catastrophic event to happen. The traditional probabilistic approach substantiates this idea by means of global predetermination of the environmental impact imposed by assessing the latter through certain probabilistic distribution. Then, assuming that the response is local, i.e. after each session of environmental impact the system returns to the same equilibrium, the probability for happening of a catastrophic event is explicitly determined by the distribution of the environmental impact. Our approach to complex systems is completely different because: (i) the hypothesis of boundedness is incompatible with the idea of thermodynamic equilibrium viewed as single state where a system resides for arbitrary long time because the concept of boundedness requires boundedness of the distance between successive states. Therefore, obviously the system must reside in a stable way in any of the states through which it passes; (ii) both parts of a complex system response, the specific homeostatic and the universal noise ones, commence from the self-organized behavior of the system. Therefore, it is to be expected that the catastrophic events actually happen gradually and thus there must be early "warning" signs for their occurrence. The question is whether these early warning signs are specific or universal. The answer to be given is that there are both specific early warning signs, considered in Chapter 2, but also there exists a universal measure for reaching certain point in the state space. Next in this Chapter we will demonstrate that this measure appears as Boltzmann-Gibbs distribution.

It will be proven that the obtained Gibbs measure has very different meaning and role than its traditional counterpart. The first one is that it is derived without prerequisite requirement for simultaneous achieving of any extreme property of the system. Thus, the derivation of the generalized form of Gibbs measure is free from imposing the condition that the equilibrium is the state of maximum entropy. Moreover, not only the obtained Gibbs measure is not an attribute of some extreme property but it serves as grounds for defining entropy! This unexpected turn is a result of the fact that neither of the traditional notions such as temperature, entropy, chemical potential is separable and falls into the well-known classes of correspondingly extensive and intensive variables. To remind, the traditional approach assumes that any system in equilibrium can be characterized by pairs of intensive-extensive variables which participate to the corresponding thermodynamical potential so that the latter to be a homogeneous function of the size of the system. Actually, the role of this split into intensive and extensive variables is to substantiate the thermodynamical limit which implies insensitivity of the thermodynamical variables to the shape and size of the system. It is obvious that the above setting is completely inappropriate for the structured systems since any structure is sensitive to the size and shape of the system. Therefore, if thermodynamical variables do exist, their participation to the thermodynamical potential must not be determined according to their association with the size of the system. That is why they must be defined by completely different considerations.

Obviously these considerations must be grounded on a number of general characteristics typical for each object regardless to whether it belongs to the class of complex systems or to the class described by the traditional statistical mechanics. At first glance the search for any universality seems a hopeless affair because the operation of each system is a result of a highly specific complex interplay among the processes that govern its internal self-organization conditioned by the corresponding constraints set by the environment. So far the study of this interplay has been focused on its particularity: usually systems are considered coupled to mass and/or heat reservoirs with features that are time-independent. An enormous research effort has been applied to the study of heat engines since the mid 19-th century onwards. Summarizing, it is firmly established that the operational characteristics of any device

are determined by a specific for the system relation between the macroscopic variables of its self-organization (e.g. concentrations) and the current environmental constraints. The variety of these relations (equation-of-state) is enormous: it ranges from the ideal gas law (Clapeyron equation) to systems of coupled ordinary/partial differential equations for far-from-equilibrium reaction-diffusion systems. The sustainability of a given operational protocol requires not only smart technology but proving robustness of its functioning to small variations of the control parameters as well. Obviously, the matter of sustainability involves also biological and social systems which are commonly known to operate at permanent variability of the environment.

At this point we suggest that the notion of equation-of-state serves as appropriate common basis for studying the macroscopic behavior of both traditional physical systems, called hereafter equilibrium systems and complex systems, called hereafter out-of-equilibrium systems. The reason is that while the equilibrium systems are characterized both by their energy function and the corresponding equation-of-state, the steady states of out-of-equilibrium systems are defined only by their dynamics, i.e. by their equations-of-state. A key advantage of this suggestion is that any universality obtained in the corresponding setting would turn equally available for both infinitely large systems and their nano-scale counterparts because it possesses the property not to pre-determine the role of the corresponding thermodynamical variables with regard to their relation to the size of the system. At the same time, we expect that the specific properties of a system, size-effects included, are concentrated in a specific steady part whose properties are described by the corresponding equation-of-state.

A successful generalization of the thermodynamics would essentially help achieving the major goal of modern nano- and informational science to meet the rapidly growing demands of the industry in an energy-effective and non-extensive way. To this end, the issue whether the behavior of the enormous variety of tailored devices exhibits common characteristics is of primary importance for developing a general strategy aimed to help resolving particular tasks. So far, the scientific community has not formed well established opinion on this topic because it still lacks a concept that, along with predicting universality, it preserves the specific-goal-oriented characteristics of a device. The major goal of the present Chapter is to develop a generalized version of certain basic thermodynamical notions so that to provide a general approach to this task.

BACKGROUND

In order to make our task clear let us first outline the major difficulties which the concept of boundedness encounters when applied to the traditional statistical mechanics.

The traditional thermodynamics is based on two fundamental laws: The first law expresses the connection between the energy conservation law and the assumption that the state variables that participate in that law do not depend on the shape and the size of the system. To make us clear, let us consider this statement in more details. The energy conservation law states that different types of energies such as internal energy, heat and mechanical energy can transform one into another so that the transformed energy to conserve:

$$\Delta U = \Delta Q + W \qquad (8.1)$$

where U is the internal energy, Q is the heat and W is the work.

Let us now remind the notion of a thermodynamical potential: thermodynamic potentials are extensive state variables of dimensions of energy. Their purpose is to allow for simple treatment of equilibrium for systems interacting with the environment. In thermodynamics all variables are either extensive or intensive. Mathematically this may be expressed in homogeneity relations with respect to the system size. Thus, extensive variables (e.g. $N, V, U, S, ...$) are first-order homogeneous functions, whereas intensive variables (like $p, T, \mu, ...$) are independent of the size of the system.

The bridge between the first law and the thermodynamical potentials is built on the assumption that there exists the so called natural variables such that by approaching the state of equilibrium they do not depend on the path. Put it in other words, the natural variables are conservative ones and therefore their differentials appear in the differential form of the first law:

$$dU = TdS - pdV + \mu dN \qquad (8.2)$$

so that S, V and N are natural variables of internal energy. With all intensive variables fixed, extensivity of all these variables means:

$$U(\lambda S, \lambda V, \lambda N) = \lambda U(S, V, N) \qquad (8.3)$$

Differentiating both sides with respect to the auxiliary parameter λ and putting $\lambda = 1$ thereafter we arrive at the identity (Euler equation for homogeneous functions):

$$U = S\left(\frac{\partial U}{\partial S}\right)_{V,N} + V\left(\frac{\partial U}{\partial V}\right)_{S,N} + N\left(\frac{\partial U}{\partial N}\right)_{S,V} \qquad (8.4)$$

From the first law it follows that

$$\left(\frac{\partial U}{\partial S}\right)_{V,N} = T\,; \left(\frac{\partial U}{\partial V}\right)_{S,N} = -p\,; \left(\frac{\partial U}{\partial N}\right)_{S,V} = \mu \qquad (8.5)$$

Thus, we arrive at the fundamental equation

$$U = TS - pV + \mu N \qquad (8.6)$$

It should be stressed on the fact that the requirement for the independence of the thermodynamical potential from the path through which a system arrives at equilibrium implies Markovianity of the transitions between different states. To remind, the Markovian property assumes that the transitions between the states are not constrained by the distance between them. Here we encounter the first fundamental difference with the concept of boundedness which explicitly constrains the admissible transitions to those, the distance to which is less than a given one. Note the difference with the Markovian property: the Markov fields also constrain the dependence of a transition to the nearest neighbors but in the *configuration* space whereas in our approach the admissible transitions are constrained in the *state* space. In other words, while the Markov fields are to be considered given *a priori*, the constraint over the admissible transitions under the concept of boundedness is "self-organized" property that commences from the functionality of a system.

The huge importance of the Second Law of the thermodynamics lies in its property to select the arrow of time for the spontaneous processes. It answers the question why the process of mixing of an ink drop into water takes place always in the direction of formation of a homogeneous solution. The traditional form of the Second Law assumes that the time arrow is selected by the processes which yield maximization of the entropy where the entropy is considered as a measure for the "disorder" in the system. More rigorously, the condition for maximization of the entropy must be considered as a condition about "forgetting" the initial conditions. This viewpoint allows assigning of a probability for a system to reside in a state of certain energy (when the energy is a conserved variable). This measure is well known as Boltzmann-Gibbs distribution. The Boltzmann distribution law says that if the energy associated with some state or condition of a system is E, then the frequency with which that state or condition occurs, is proportional to:

$$\exp\left(-\frac{E}{kT}\right) \tag{8.7}$$

where T is the system's absolute temperature and where k is the Boltzmann constant.

However, this scheme is inappropriate for complex systems because of the following reasons: there are two major controversies that feature the conflict between the behavior of the complex systems and the traditional thermodynamics:(i) what determines the stability of a system and how the system behaves so that to sustain stability; (ii) how much work can be extracted from a system.

Let us start with the first controversy. In the traditional thermodynamics we identify and measure as the thermodynamic energy U of a macroscopic system its total mechanical energy E: the sum total of all the kinetic and potential energies of all the molecules that make up the system. Two key observations lead to the present understanding. The energy E of an isolated mechanical system is a constant of the motion; although the coordinates and velocities of its constituent parts may change with time, that function of them that is the energy has a fixed value, E. That is at the mechanical level. At the thermodynamic level, as one aspect of the first law of thermodynamics, it is recognized that if a system is thermally and mechanically isolated from its surroundings – thermally isolated so that no heat is exchanged ($q = 0$) and mechanically isolated so that no work is done ($w = 0$) – then the function U of its thermodynamic state does not change. That is one fundamental property that the mechanical E and the thermodynamic U have in common. The second is that if the mechanical system is not isolated, its energy E is not a constant of the motion, but can change, and does so by an amount equal to the work done on the system: $E = w$. Likewise, in thermodynamics, if a system remains thermally insulated ($q = 0$), but is mechanically coupled to its environment, which does work w on it, then its energy U changes by an amount equal to that work: $U = w$. This coincidence of two such fundamental properties is what led to the hypothesis that the thermodynamic function U in the first law of thermodynamics is just the mechanical energy E of a system of some huge number of degrees of freedom: the total of the kinetic and potential energies of the molecules. If our system is not isolated but is in a thermostat that fixes its temperature T and with which it can exchange energy, then the energy E is not strictly constant, but can fluctuate. Such energy fluctuations in a system of fixed temperature, while often interesting and sometimes important, are of no thermodynamic consequence: the fluctuations in the energy are minute compared with the total and are indiscernible at a macroscopic level. Therefore the thermodynamic energy U in a system of fixed temperature T may be identified with the mean mechanical energy $\bar{E}$ about which the system's mechanical energy fluctuates.

The plausibility of the above considerations is explicitly grounded on the specific pre-determination of the environment: the system is either isolated, or coupled to heat reservoir of constant intensity etc. Therefore, these considerations turn problematic when applied to complex systems set in an ever-changing environment. Note that the major difference between the traditional statistical mechanics and the concept of boundedness is that while the former *a priori* sets a specific environment of constant intensity (coupling to heat and mass reservoirs of constant intensity) the concept of boundedness implies neither specific environment nor the constancy of the intensity of its impact. Moreover, it considers the environment acting in non-linear and non-homogeneous way on a system. The only constraint imposed is that of boundedness. Thus, it is impossible to define integrals of motion and/or other conserved quantities. Consequently, the entire traditional setting for derivation of the Boltzmann-Gibbs distribution turns inappropriate. Moreover, it makes impossible

to associate the internal energy with the mechanical energy of the system.

Further, we face two other major conflict issues between the concept of boundedness and the traditional statistical mechanics:

1. The idea about the independence of the arrival at equilibrium from the path is apparently challenged by the fact that the admissible transitions in the state space are bounded by the functionality of the system. It is worth noting that the constraint over the admissible transitions implies their non-Markovianity. This consideration gives rise to the question whether the substantiated by the non-Markovianity dependence of the final state on the path implies violation of the notion of equilibrium. In other words, this conflict poses the question what is the appropriate characterization of a system which has no integrals of motion?

2. The traditional statistical mechanics tacitly assumes that a system exerts work without undergoing any structural changes. This is the case when a system is structureless, an ideal gas for example, but is very important when considering systems of self-coordinated functionality. Then, the involvement/extraction of any amount of matter/energy could cause either functional or structural rearrangement of the system. Then, what would we call "work" that the system does?

So far, we pose a number of questions the most important of which are: Does the proposed in the Introduction idea about the use of the equation-of-state, viewed as common characteristics of the traditional and complex systems, serves as a well substantiated premise for developing of the new approach? The next question is whether or not the grounding assumption for the derivation of the generalized analog of Boltzmann-Gibbs measure relates to the hierarchical organization of the semantic-like response?

RESPONSE TO ENVIRONMENTAL VARIATIONS MODELED BY A BIS

The grounding assumption of our approach is the assertion that the ability for functional self-organization is the major property that provides stable behavior of a system. This assertion is fundamentally different to the major assumption of the traditional thermodynamics which supposes that in constant environment a system eventually reaches equilibrium. Note that our statement does not pre-suppose any conditions on the environment other than those associated with providing and sustaining the stability of a system. Thus, our approach is open to be applied to an ever-changing environment. The only condition imposed on the variability of the environment is the boundedness. To remind, it implies that the succession of the environmental impacts is kept within given margins so that the corresponding response not to exceed the thresholds of stability of a system; the rate of exchanging matter/energy between the system and its environment is also kept bounded. Alongside, we assume that the response is local and is determined by the current local impact and the current state of the system. Thus, we tacitly presuppose that a system could stably reside in more than one state (equilibrium). Further, we assume that each of these states is characterized by its equation-of-state, i.e. a specific for the state and the system relation between the parameters of the environmental impact and the self-organization of the system in response to that impact.

Actually, the equation-of-state is well established way of characterization of the operational characteristics of any device determined as specific for a system's relation between the macroscopic variables of its self-organization (e.g. concentrations) and the current environmental constraints. The variety of these relations (equations-of-state) is enormous: it ranges from the ideal gas law (Clapeyron equation) to systems of coupled ordinary/partial differential equations for far-from-equilibrium reaction-diffusion systems.

The sustainability of a given operational protocol requires not only smart technology but proving robustness of its functioning to small variations of the control parameters as well. Obviously, the matter of sustainability involves also biological and social systems which are commonly known to operate at permanent variability of the environment. This substantiates the link between the academic issue about the generalization of the thermodynamics so that to be available for a variable environment and the practical matter of sustainability of natural and tailored devices of different utility.

At this point we suggest that the notion of equation-of-state serves as appropriate common basis for studying the macroscopic behavior of both equilibrium and out-of-equilibrium systems. The reason is that while the equilibrium systems are characterized both by their energy function and the corresponding equation-of-state, the steady states of out-of-equilibrium systems are defined only by their dynamics, i.e. by their equations-of-state. A key advantage of this suggestion is that any universality obtained in the corresponding setting would turn equally available for both infinitely large systems and their nano-scale counterparts. At the same time, we expect that the specific properties of a system, size-effects included, are concentrated in a specific steady "homeostasis" whose properties are described by the corresponding equation-of-state.

Next we will prove that the response of each system to a fluctuating environment modeled by a bounded irregular sequence (BIS) is additively decomposable to a steady specific component (homeostasis) whose properties are described by the corresponding equation-of-state and a stochastic one which has universal characteristics.

The enormous variety of systems and interactions system-environment implies that any universality is to be derived on the grounds of a few weak constraints only. We claim that a single constraint turns sufficient and this is the constraint of boundedness. It implies that environmental variations succeed in arbitrary order but so that:

1. Their amplitude remains bounded within specific to the system margins;
2. The increments are also bounded – this constraint naturally arises from the condition that to stay stable, a system exchanges only finite amount of energy/matter/information with the environment;
3. The rate of their succession is to be bounded so that a system to be able to reach steady state at current control parameters.

Note that the boundedness depends neither on the dimensionality of a system nor on its symmetry – it is applied to each control parameter and to each dimension separately.

We model a fluctuating environment by a vector of control parameters each component of which is a BIS. Then, the most general form of an equation-of-state in a fluctuating environment reads:

$$\vec{F}\left(\lambda, \vec{n}, \frac{\partial \vec{n}}{\partial t}, \frac{\partial \vec{n}}{\partial x_i}, \frac{\partial^2 \vec{n}}{\partial x_i \partial x_j}\right) = 0 \qquad (8.8)$$

where $\vec{\lambda} = \vec{\lambda}_0 + \delta\vec{\lambda}_{st}$ is vector of the control parameters separated according to Lindeberg theorem (Feller, 1970) into expectation value λ_0 and stochastic component $\delta\vec{\lambda}_{st}$; $\vec{n}$ is the vector of the macroscopic variables associated with the organization of the system (e.g. concentrations); the derivatives follow their traditional meaning (time and space correspondingly).

Equation (8.8) determines the system response $\vec{n}\left(\vec{\lambda}_0 + \delta\vec{\lambda}_{st}\right)$ to environmental variations which appears as a stochastic sequence whose terms are specified by the solution of Equation (8.8) for each realization of $\delta\vec{\lambda}_{st}$. In general, the equation-of-state Equation (8.8) may be arbitrary (alge-

braic or differential; deterministic or stochastic; linear or non-linear) - in each particular case its concrete form is specified by the interaction between a given system and its environment. It is subject to the following common constraint: it must be such that the stochastic sequence $\vec{n}\left(\vec{\lambda}_0 + \delta\vec{\lambda}_{st}\right)$ is also a BIS. This condition is necessary for meeting the natural requirement that a system stays stable if and only if a finite amount of energy/matter/information is exchanged with the environment in every moment. Therefore, the response $\vec{n}\left(\vec{\lambda}_0 + \delta\vec{\lambda}_{st}\right)$ is also a BIS and hence it is decomposable to expectation value (steady component) and a stochastic component:

$$\vec{n}\left(\vec{\lambda}_0 + \delta\vec{\lambda}_{st}\right) = \vec{n}_{\exp} + \delta\vec{n}_{st} \qquad (8.9)$$

where $\vec{n}_{\exp}$ is the expectation value and $\delta\vec{n}_{st}$ is a zero-mean BIS.

It should be stressed that the expectation value $\vec{n}_{\exp}$ from Equation (8.9) is not equal to $\vec{n}\left(\vec{\lambda}_0\right)$! This happens because the non-linearity of the equation-of-state Equation.(8.8) renders the Taylor expansion of the terms in the stochastic sequence $\vec{n}\left(\vec{\lambda}_0 + \delta\vec{\lambda}_{st}\right)$ around $\vec{\lambda}_0$ inappropriate. Yet, since $\vec{n}\left(\vec{\lambda}_0 + \delta\vec{\lambda}_{st}\right)$ is BIS each term of which is determined by Equation (8.8), the expectation value $\vec{n}_{\exp}$ can be presented as a solution of the equation-of-state Equation (8.8) but at shifted control parameters $\vec{\lambda}_s$:

$$\vec{F}\left(\vec{\lambda}_s, \vec{n}_{\exp}, \frac{\partial \vec{n}_{\exp}}{\partial t}, \frac{\partial \vec{n}_{\exp}}{\partial x_i}, \frac{\partial^2 \vec{n}_{\exp}}{\partial x_i \partial x_j}\right) = 0 \quad (8.10)$$

The shift between $\vec{\lambda}_s$ and λ_0 is highly specific to the system and the control parameters!

Thus, Equation (8.10) appears as the equation-of-state for the steady component (homeostasis) of the response. It is worth noting that since it preserves the structure and the properties of Equation (8.8), the homeostasis preserves the specific goal-oriented properties of the system considered. Since the above decomposition is universal, i.e. it holds for arbitrary equation-of-state and any environment, subject only to boundedness, the enormous variety of systems opens the door to realization of a great variety of behavioral patterns ranging from a stationary state to a limit cycle, a self-assembled structure, morphological organization etc.

Outlining, one concludes that under the very weak constraint of boundedness alone, the response of a system to a varying environment, modeled by a BIS, is additively decomposable into a specific steady component whose properties are defined by the corresponding equation-of-state, and a stochastic one which has universal characteristics. The decomposition is a generic property of Equation (8.8) and is insensitive to the nature of the system, to the particularities of its interaction with the environment, and to the statistics of the environmental variations.

It is worth noting two things. The first one is the additive decomposability of the equation-of-state to a specific and universal part so that the specific part to be kept robust to small environmental variations whereas the universal part to exhibit properties insensitive to the statistics of the environmental fluctuations. Thus, the above considerations substantiate the first fundamentally different property of the present approach compared to the traditional thermodynamics: we replace the notion of equilibrium with the notion of homeostasis. The major difference between both notions is that while the thermodynamical equilibrium is a state which is supposed to be a global one, i.e. it is a single state which is a global attractor for all initial conditions, the homeostasis is a state whose robustness holds in a certain domain of control parameters. The limited robustness turns rather advantageous because our approach open the door for strengthening

of the response by means of "adaptation" of the homeostasis. The latter implies partitioning of the control parameters space into basins-of-attraction where, although all of them are governed by the same equation-of-state, the spatio-temporal pattern which presented homeostasis is different. Thus, by means of "adapting" its homeostasis to the changes in the environment, the systems "enlarge" the domain of their stable behavior.

The second major distinction between our approach and the traditional thermodynamics is the highly non-trivial relation between the control parameters characterizing the current environment and the effective parameters which governs the equation-of-state for the corresponding homeostasis (see considerations after Equation (8.9)). Thus, the highly non-trivial interplay between a system and the environment once again implies that the system response is active. To compare, the traditional statistical thermodynamics model the environment impact by certain *a priori* set distribution and, on the grounds of the assumption that the response is proportional to the impact, actually suggests judging about the impact by studying the response. However, our approach justifies that the impact essentially contributes to the self-organization of the response which makes the relation impact-response highly non-linear and non-homogeneous.

Generalized Gibbs Measure

We have already established that the steady component of the response is specific to a system and robust to small environmental fluctuation and because of these properties we called it a homeostatic component. Yet, now we face the question about the properties of the stochastic component. Our major aim is to prove that it provides the universal property that the probability a system to reside in a certain state is given by a generalized form of Gibbs measure. The proof of its universality is explicitly grounded on the property of the stochastic component to be a BIS. In Chapter 1 we have proved that BIS`s have certain characteristics that appear as time series invariants. These findings will help us to answer the question about the measure for a system to reside in a given state. The issue about measure for residing in a given state is provoked by the non-linearity of the response which renders possible that certain members of the stochastic sequence $\delta\vec{n}_{st}$ to provoke departure from that given state. Intuitively, it seems that the solution is very simple: for any particular sequence the minimal requirement is that one can calculate the probability for $\delta\vec{n}_{st}$ to exceed the corresponding threshold. Hence, no universality is to be expected. However, this approach suffers a serious drawback: it does not take into account the role of the dynamics of the environmental variations. Its non-triviality is best revealed by the obtained in Chapter 1 result that the state space of a BIS is a dense transitive set of orbits which implies that the motion in the attractor is indecomposable. Put it in other words, the latter implies that the average probability for escape is a non-local event and thus it depends not only on the current amplitude of the environmental variations but also on the dynamics of their succession, and on the succession of states through which the system passes. Put it in other words, the non-locality of the escape is a result of the boundedness: the changes happen gradually and the system is not supposed to return to a single equilibrium state after each session of impact. Next we will demonstrate that the complex interplay between the local amplitude of the environmental variations and the non-locality of their dynamics renders the average probability for being at a given state to be described in a universal form. Further, it turns out that the non-locality of the dynamics is explicitly related to the heat dissipated by fluctuations and thus provides it`s appearing as an effective temperature in the expression for the average probability for residing in a given state. On the other hand, the boundedness imposed on the amplitude of fluc-

tuations makes available to consider the orbital motion as confinement by a potential. We prove that the balance between the potential confinement and the random motion that comes from the dynamics of the fluctuations renders the average probability for residing at a given state to appear in a universal form which generalizes the notion of the Boltzmann-Gibbs weight. Next follows the proof.

The structure of a BIS state space viewed as a dense transitive set of orbits implies that the motion in it is completed as a random walk of blobs superimposed on an orbit where blobs are orbits on finer scale. The orbital motion on a coarse-grained scale is due to boundedness imposed on the amplitude of the variations. Accordingly, it acts as a confinement by a "potential" whose characteristics depend on the characteristics of the given orbit. On the other hand, the motion on the finer scales is random and is specified by the corresponding diffusion coefficient. Assuming that the diffusion coefficient exists, a statement that we shall prove later, it is obvious that the stationary motion in the state space is provided by the following balance between the random motion on finer scales and the "potential" confinement of the orbits on a coarse-grained scale:

$$h\bar{D}_U \frac{\partial P}{\partial n} = -\frac{\partial U(n)}{\partial n} P \qquad (8.11)$$

where P is the probability for residing on an orbit whose "potential" is $U(n)$; the properties of $U(n)$ are derived by means of considering the orbit as a Fourier knot, n is a point is the state space (it is worth noting that the points in the state space represent the values of the macroscopic variables not positions and velocities of the species in the system!); $\bar{D}_U$ is the average diffusion coefficient and h is a constant. Equation (8.11) is always one-dimensional because the boundedness is insensitive to the symmetry of the system and its state space, and its dimensionality. Then its solution reads:

$$P(n) \propto \exp\left(-\frac{U(n)}{h\bar{D}_U}\right) \qquad (8.12)$$

Next item is to demonstrate that $h\bar{D}_U$ has the meaning of an effective temperature. For this purpose let us first derive an explicit formula for the average diffusion coefficient of a BIS bounded by a threshold corresponding to "potential" $U(n)$. The definition of a diffusion coefficient reads:

$$\bar{D} = \lim_{t \to \infty} \frac{X^2(t)}{t} \qquad (8.13)$$

where $X(t)$ is a deviation of a trajectory from the expectation value. In Chapter 5 we have established that the deviations from the expectation value of a BIS are developed as excursions of certain amplitude and duration; the relation between them is set by a power law where the exponent is specific to the system:

$$X \propto \Delta^{\beta(\Delta)} \qquad (8.14)$$

where X is the amplitude of an excursion; Δ is the duration of its development; $\beta(\Delta)$ is an exponent set by the particularities of fine scale dynamics.

Further in Chapter 5, we have established that the distribution of excursions reads:

$$G(X) = cX^{1/\beta(X)} \frac{\exp\left(-X^2/\sigma_U^2\right)}{\sigma_U} \qquad (8.15)$$

where σ_U is the variance of the sequence.

Then, Equation (8.13) becomes:

$$\bar{D}_U = \lim_{t \to \infty} \frac{\int_0^{X_U} X^2 G(X)}{\int_0^{\Delta_U} \Delta G(\Delta)} \qquad (8.16)$$

By means of simple calculations $\bar{D}_U$ becomes:

$$\bar{D}_U \propto \frac{(\sigma_U)^2}{(X_U)^\alpha} \qquad (8.17)$$

where $\alpha \ll 1$. Thus, Equation (8.17) proves that the diffusion coefficient of the random walk in the state space of a BIS is proportional to its variance. On the other hand, we have proved in Chapter 1 that the integrated power spectrum of a BIS is among its time series invariants and is also proportional to its variance. The non-triviality of this coincidence is revealed by the Nyquist theorem (Gardiner, 1985) which states that the integrated power spectrum is the measure for the heat dissipated in the interaction system-environment. Thus, one concludes that the diffusion coefficient D_U is to be considered proportional to an effective temperature whose measure is the variance of the time series. Then, Equation (8.12) becomes:

$$P(n) \propto \exp\left(-\frac{U(n)}{kT_{eff}}\right) \qquad (8.18)$$

It is worth noting that $P(n)$ from Equation (8.18) appears not only as a measure for robustness to environmental variations but at the same time it serves as local thermodynamical potential for the given state of the system. More precisely, it appears as local Gibbs measure for a stationary bounded random motion. A crucial property of $P(n)$ is that it is not derived on the condition of entropy maximization but on the condition of balance between random and orbital motion given by Equation (8.11). To remind that in the traditional statistical mechanics the Boltzmann-Gibbs weight is that distribution whose entropy is maximal among all distributions that share the same fixed average energy. It is worth noting, however, that this derivation is set by the explicit use of energy function as the basic characteristic of a system. However, as already have been stated above, the energy function is not appropriate for characterizing the out-of-equilibrium systems. In order to overcome this difficulty, our derivation of the Boltzmann-Gibbs measure is set on the grounds of the equation-of-state and thus it turns equally available for both equilibrium and out-of-equilibrium systems. Its major advantage is that now it is available not only for simple physical systems but it is available also for systems that involve macroscopic self-organization including pattern formation, hierarchical organization etc, i.e. for systems that apparently violate the entropy maximization as a condition for equilibrium.

It is worth noting that the balance between orbital and random motion expressed through Equation (8.11) implies a ban over spontaneously aroused fluxes. Thus, it is in accordance with the condition that the semantic trajectories are those trajectories where the action is stationary. Let us remind that the stationarity of the action substantiates an inter-level feedback which "imposes" hierarchical relations "top down," i.e. it imposes links from higher levels of the hierarchical self-organization to the lower ones. Thereby the Boltzmann-Gibbs measure serves as a ban over "self-propelled" semantics: it is impossible to "generate" a semantic-like response from the "noise" (random motion) only; the necessary ingredient for execution of the semantics is the hierarchical self-organization. This implies that the semantic-like response appears as a result of the coherent action of hierarchical order both bottom up and top down implemented correspondingly by the boundedness over the rates in the direction

bottom up and by the stationarity of action top down. The major goal of this coordination between the hierarchical levels, and consequently the organization of their response in a semantic-like way, is to provide long-term stability of the system by means of not allowing sporadic and accidental non-coordinated "reconstructions" inside the system. To remind, the sporadic "reconstructions" are automatically "damped down" by means of matter wave emitting targeting at releasing of the extra-accumulated matter/energy.

Accordingly, the notion of entropy acquires novel understanding. The above considerations open the door for defining the entropy as specific for a system non-probabilistic thermodynamic function that describes the amount of energy necessary for adapting the self-organization of that system to ever-changing control parameters. Indeed, the presence of a thermodynamical potential (Gibbs measure) renders the explicit mathematical expression of the entropy to be defined on the grounds of the Maxwell relations so that its specification for each particular system is rendered by the corresponding equation-of-state. Thus, the present approach eliminates the major obstacle to the development of the notion of entropy beyond the simple physical systems, that of entropy maximization.

It is worth noting that the obtained by us Gibbs measure is substantiated by the equation-of-state because it is brought about by the balance between the fine-grained random motion and coarse-grained orbital one where the relation between them commences from the corresponding equation-of-motion. Thus, since the Gibbs measure has the meaning of the internal energy, the relation between $U(n)$ and the equation-of-state relates the internal energy and the functionality of a system in a non-extensive way. It should be stressed that unlike the traditional thermodynamics which uses the first law of the thermodynamics to define explicitly the notion of the internal energy, the proposed approach provides an autonomous way of defining the internal energy which is explicitly grounded on the functional properties of the system. It should be stressed also that the notion of "temperature" in our approach is also explicitly related to the functional properties of the system and is expressed in a non-extensive way through the variance of the "noise" part of the power spectrum.

Further, it turns out that the present approach provides unified thermodynamical description not only for both equilibrium and out-of-equilibrium systems regardless of their size and level of macroscopic self-organization but it also brings about a completely novel prospective on the notion of information. Certainly, the release from the condition of entropy maximization makes available the association of different logical units with the steady states that are carried out at different control parameters, and the non-probabilistic definition of entropy provides the information encapsulated in each logical unit to be characterized by the specific for each of those steady states properties not necessarily related to the entropy. *This setting is fundamentally different from the Shannon probabilistic approach to information which, on the contrary, leaves the notion of information insensitive to the specific properties of the steady states and their dynamics.*

CHEMICAL POTENTIAL

The major aim of the present Chapter is to prove that the basic notions of thermodynamics acquire new understanding in the frame of the concept of boundedness. The fundamental difference between the traditional thermodynamics and the concept of boundeness is that the former sets *apriori* the environment whereas the later lets the environment free of any specification other than the mild condition of boundedness alone. This setting provides the major difference between the both approaches: while the traditional thermodynamics is grounded on the assumption that the basic characteristics of a system, internal energy, entropy, chemical potential are functions of the state only and do not depend on the path through which a system arrives at a given state, the concept of boundedness suggests that although the above characteristics are functions of the state alone, their changes from a state to a state are always finite (this means that they cannot be made infinitesimally small) and thus they depend on the path through which a system arrives at a given state. This difference constitutes the new frame within which these characteristics must be derived: thus, unlike the traditional approach which utilizes the use of extreme properties such as maximization of entropy at fixed total energy, the present approach is grounded on the utilization of the universal properties of a state space under boundedness derived in the present book.

In the previous section we derived the internal energy and the temperature on the grounds of the properties of the state space under boundedness. The goal of the present section is to continue with the derivation of other crucial notion of the thermodynamics: that of the chemical potential. Its most widespread definitions reads that it is the Gibbs energy necessary for associating or dissociating a particle in/out a system. In the traditional thermodynamics the chemical potential is supposed to be an intensive variable, i.e. it must be independent from the number of species. However, next we will present a proof that the chemical equilibrium cannot sustain this property of the chemical potential. Thus, it turns out that the traditional approach suffers serious drawback. Before considering the new definition of the chemical potential we present a detailed explanation of this inherent controversy of the traditional thermodynamics.

Let us follow the conventional derivation of the necessary conditions that a chemical reaction at fixed pressure and temperature should obey. We will demonstrate that it yields violation of the condition for the chemical potential to be an intensive variable. In equilibrium the amount of each of the reactants and products is constant and corresponds to the stoicheometric relation:

$$\sum_{i=1}^{S} \nu_i A_i = 0 \qquad (8.19)$$

where ν_i is the stoicheometric coefficient of the $i - th$ sort, A_i is its amount and S is the number of the $i - th$ sort entities.

Equation (8.19) is equivalent to the following relation:

$$\frac{\partial \Phi}{\partial N_i} + \frac{\partial \Phi}{\partial N_2} \frac{\partial N_2}{\partial N_1} + ... + \frac{\partial \Phi}{\partial N_i} \frac{\partial N_i}{\partial N_1} + ... + \frac{\partial \Phi}{\partial N_S} \frac{\partial N_S}{\partial N_1}$$
$$= 0$$
$$\qquad (8.20)$$

where Φ is the Gibbs energy, N_i is the number of entities of the $i - th$ sort.

Let us now rewrite (8.20) in slightly different form:

$$\sum_{i=1}^{S} \frac{\partial \Phi}{\partial N_i} \frac{\nu_i}{\nu} = 0 \ . \qquad (8.21)$$

By definition:

$$\frac{\partial \Phi}{\partial N_i} = \mu_i \qquad (8.22)$$

where μ_i is the chemical potential.

Thus, at equilibrium the following relation holds:

$$\sum_{i=1}^{S} \nu_i \mu_i = 0 \qquad (8.23)$$

Further, since the Gibbs energy is thermodynamical potential in the so-called linear domain around the equilibrium state, i.e. in its the nearest neighborhood, the condition for its smoothness:

$$\frac{\partial^2 \Phi}{\partial N_i \partial N_j} = \frac{\partial^2 \Phi}{\partial N_j \partial N_i} \qquad (8.24)$$

leads to the following relation:

$$\frac{\partial \mu_i}{\partial N_j} = \frac{\partial \mu_j}{\partial N_i} \qquad (8.25)$$

where μ_i, μ_j are the chemical potentials of the $i - th$ and $j - th$ sort entities correspondingly. Let us now have a closer look on (8.25) - it turns out that (8.25) makes the chemical identity dependent on the number of the entities! In addition, it is obvious that thereby the chemical potential turns a non-intensive variable so that it depends not only on the number of the species of a given sort but on the number of species of all other sorts of species. However, according to its meaning (8.22), the chemical potential is predetermined by the chemical identity of the entities. That is why, it is hardly to be expected that it would depend on the current number of the entities of the other types in particular for ideal gases where no other interactions except for the collisions are present. If so, the medieval alchemist dream would have come true - the chemical identity changes under the change of the number of entities!

So, obviously the definition of the chemical potential needs serious revision. Next we propose new definition associated with the properties of the state space under boundedness. The advantage is that it is straightforwardly related to the issue of stability. The fundamental reason to associate the notion of the chemical potential with the stability is the complexity of the interaction between any system and its environment that can entangle them so that sometimes their individuality is blurred. That is why it is important that the notion of the chemical potential involves the functional relations introduced by the interaction between the system and the environment.

A crucial test for the new definition of the chemical potential is whether it puts a ban over association of arbitrary amount of species. This is very important for delineating a system from its environment in the process of their interaction under unspecified environmental conditions. Thus, if the exchange of species is not limited, a system could either enlarge to arbitrary size or deplete to extinction. Further, the same problem arises when considering formation of local fluctuations in a solution for example. To make us clear let us consider as an example the Brownian motion. It is accepted that the random walk of each pollen is a result of a random "kick" that comes from the solvent. It is assumed that the "kicks" are result of the action of random "sources" that originate from the discrete atomic structure of the solvent. The intensity of the random sources is supposed to vary in an infinite range. Then the random walk of each pollen is successfully described by adding stochastic force to the deterministic friction in the equation of motion so that the stochastic force to match the behavior of the random "kicks." This equation is called Langvin equation. Further, the transition from the Langvin to the Einstein-Smoluhowsky equation for the diffusion on coarse-grained time scale demands a well-defined concentration of the pollen species. However, this requirement immediately demands finite spatial size and finite amplitude of the fluctuations, random "kicks" included. So, it poses the question that there must be a mechanism that guarantees destruction of arbitrarily large fluctuations. On the other hand, since the size of a pollen is much larger than the size of solvent molecules, it is obvious that the "sources" must be spatially extended. Thus, the mechanism that destroys the large fluctuations should be "flexible"-it should select "good" small fluctuations and destroy the "bad" large ones.

The current status-quo cannot help because the chemical potential in the thermodynamics accounts only for the chemical identity of a single entity and allows associating of arbitrarily large number of entities which immediately allows growing of a fluctuation to arbitrarily large size. Next we advance a new definition of the chemical potential that involves functional relations among entities and relates the number of associated entities with the stability of the system. One of the major outcomes of that definition is that it always provides finite size and finite lifetime of extended fluctuations.

Chemical Potential under Boundedness

Turning point of the developed by us generalization of the thermodynamics is the definition of the chemical potential so that to involve the functional relations among the entities and the factors which provide long-term stability. The reason is twofold: on the one hand the interacting systems in the reality are entangled by the processes that proceed among them. Hence, sharp separation between a system and environment is impossible which makes important to define the chemical potential so that to involve the functional relations created in the process of interaction. Furthermore, since the functional relations are not constant in the course of time, their modifications changes the chemical potential and can make it to turn to zero. The latter, however, implies collapse of the system. This calls for explicit relation between the chemical potential and the stability of a system. On the other hand, the traditional definition of the chemical potential does not consider stability under associating/dissociating entities which makes possible associating/dissociating of arbitrary number of entities without any effect on the stability of the system. In order to involve the stability we define the chemical potential through the properties of the state space:

$$\mu_i = -\Omega \frac{\delta L}{\delta n_i} \qquad (8.26)$$

where Ω is the volume of the system; n_i is the number of the entities of the $i-th$ sort; L is the Langrangian of a system. Equation (8.26) describes a process initiated by small deviations from the general equilibrium condition $\delta S = 0$, where S is the action. The advantage of this definition is that it is not related to the thermodynamics but has deeper origin, namely: since the functional relations in a many-body system are built-in in the state space, Equation (8.26) is an explicit expression for taking into account the intrinsic "relations" among the constituting entities as well as their modification under the interaction of the system with its environment. By means of the new definition we will demonstrate that a system stays stable if and only if the amount of energy/matter that it exchanges with the environment is bounded.

Since the action and the Lagrangian are explicitly related to the motion in a state space, it is to be expected that the chemical potential is also explicitly related to the properties of the state space. It should be stressed that among these properties, the present derivation meets the stationarity of the action. Thus, the condition for stationarity of the action appears as common ground for our generalized thermodynamics: to remind that the derivation of the Gibbs energy and the effective temperature considered in the previous section are grounded also on the condition for stationarity of the action.

The further considerations are grounded on the explicit relation between the notion of chemical potential and stability of the system. Indeed, the chemical potential must be a measure how strong the stability of the system is "affected" by association or dissociation of an entity. To compare, the traditional thermodynamics leaves the chemical potential intact during the interaction. In turn, this allows a system to associate/dissociate arbitrary number of entities without any affect on its stability. In the Chapter 5 it has been found out that the property of stretching and folding of the state

space trajectories is interrelated with the stability of the system - whenever certain relation among 3 general characteristics of the stretching and folding holds, the system is asymptotically stable (Equation (5.13)). The question now is how the stability is related to the problem of associating and dissociating of entities.

The general condition for breaking the stability of a system is turning its chemical potential to zero. Let us now come back to the structure of the state space under boundedness: in general it can be separated to a "bulk" and a "surface" part. The former one is associated with the core of the state space where (5.13) holds in any direction. Therefore, the chemical potential of the "bulk" part is not affected by the number of exchanging entities; it is rather to be associated with that part of the system that remains stable under an interaction. On the contrary, the "surface" part is associated with the current boundary of a state space under boundedness; in other words it is to be associated with the interaction with the environment. The corresponding part of the chemical potential is strongly sensitive to the current curvature that in turn strongly depends on the current number of exchanged entities. Now we will present an example how the total chemical potential turns to zero under the exchange of finite number of entities. Since any exchange of an entity modifies the shape of state space "surface," the natural measure of the surface part of the chemical potential μ_s is its local curvature:

$$\mu_s = \int_S \alpha k ds \qquad (8.27)$$

where α is the density of the surface energy; S is the area of the state space surface. The permanent variations of the number of entities result in permanent modification of the value and sign of the local curvature. So, μ_s permanently varies and eventually turns the total chemical potential μ_{tot} to zero. This immediately yields the system falling apart. The next task is to illustrate that this happens at finite values of the state space variables, i.e. when the number of exchanged entities is limited. According to the above considerations the destruction of the system happens whenever:

$$\mu_{tot} = 0 \tag{8.28}$$

Since the bulk part of the chemical potential is insensitive to the variations of the state variables, Equation (8.28) holds whenever the following relation holds:

$$\mu_s = -\mu_b \tag{8.29}$$

where μ_b is considered *a priori* set constant.

Hereafter our task is to illustrate that the present definition of the chemical potential indeed involves functional relations that appear in any process. Let us consider a chemical reaction that selects two relevant sorts of entities: their numbers are denoted by x and y correspondingly. The state space is two-dimensional and its "surface" can be parameterized as follows:

$$\begin{aligned} x &= r^{a(\theta)} \cos \theta \\ y &= r^{b(\theta)} \sin \theta \end{aligned} \tag{8.30}$$

where the powers $a(\theta)$ and $b(\theta)$ comprise the permanent change of the local curvature through the dependence $\theta = \theta(t)$. Then the current local curvature k reads:

$$k = \frac{\left(\dot{x}^2 + \dot{y}^2 \right)^{3/2}}{\dot{x}\ddot{y} - \dot{y}\ddot{x}} \tag{8.31}$$

where the derivation is with respect to time. Simple algebraic calculations yield:

$$k = \frac{1}{r^{2a-b}} \frac{1 - A - \dfrac{B}{r}}{\left(\sin^2 \theta + r^{2(b-a)} \cos^2 \theta \right)^{3/2}} \tag{8.32}$$

where:

$$A = \dot{a}ab^2b\left(b-1\right)\sin^2 \theta - \dot{b}b\dot{a}^2a\left(a-1\right)\cos^2 \theta \tag{8.33}$$

$$B = \left(\dot{b}^2 + b\ddot{b} \right)\sin^2 \theta + \left(\dot{a}^2 + a\ddot{a} \right)\cos^2 \theta \tag{8.34}$$

Since a, b and their derivatives permanently vary, due time course Equation (8.32)-Equation (8.34) select finite $\vec{r}_{cr} = \left(x_{cr}, y_{cr} \right)$ so that Equation (8.28)- Equation (8.29) are satisfied. The values of a, b, their derivatives and μ_b are set on the particularities of the reaction kinetics. Evidently, the system stays stable until its state variables remain bounded so that:

$$\begin{aligned} x &< x_{cr} \\ y &< y_{cr} \end{aligned} \tag{8.25}$$

where we called $\vec{r}_{cr} = \left(x_{cr}, y_{cr} \right)$ kinetic threshold of stability. It should be stressed that (8.25) is to certain extend analogical to Equation (5.13) even though not equivalent. The kinetic threshold of stability selects the largest possible size so that the system in question remains stable under the exchange of matter/energy with the environment. Then Equation (8.25) can be viewed as a necessary condition for a system to stay permanently stable in the process of an interaction. Besides, since the fluctuations develop also through exchange with the environment, it can be viewed likewise as necessary condition for limiting their spatial size, amplitude and lifetime. On reaching the critical size, all fluctuations that initially have smaller size are destroyed; those of larger size cannot be created at all.

Now we are able to explain why only small size fluctuations with finite lifetime are developed at Brownian motion. Let us remind the considerations made in the beginning of the section about the pollen random walks under the "sources" whose origin is the discrete structure of the solvent - these random "sources," viewed as fluctuations, should have finite size and finite lifetime so that to provide coherent "efforts" of the water molecules in order to move randomly and independently from one another the "gigantic" pollens. So, the coherent "efforts" are provided by those fluctuations whose size does not exceed the kinetic threshold of stability; to remind that it is impossible to create larger ones. The finite lifetime appears in the course of the interaction with either the pollens or the other fluctuations: sooner or later the number of exchanged molecules will exceed the kinetic threshold of stability which will result in the collapse of the corresponding fluctuation.

It should be stressed that the spatial boundedness of fluctuations is insensitive to the dimensionality of the system because it is set on dimensionality of the state space which is not proportional to the spatial dimension!

Efficiency of Nano-Particles

One of the major results of our approach to the behavior of complex systems in a non-predetermined environment is the ubiquity of macroscopic fluctuations. Taking into account their presence is of fundamental importance for the technology. Next we will consider certain outcomes of our approach which are relevant from the practical point of view and in particular for the nano-technology.

Let us start with considering the effect of the development of macroscopic fluctuations in large industrial reactors. Usually, the industry reactors are constructed so that reaction to proceed at steady stable regime. However, the development of fluctuations seriously perplexes the situation. Indeed, fluctuations of large enough size could give rise to undesirable effects such as phase transitions, change of the dynamical regime etc. Hence an indispensable part of the reactor construction must be the development of appropriate feedback so that to prevent those fluctuations. However, this matter is a complicate task that requires a lot of efforts and elaborate mind. That is why the question is whether there is another way out.

There is another way out and it is straightforwardly related to the modern nano-technology. One of the aims of the nano-technology is a deep miniaturization so that to have more effective use at less costs. Applied to the catalysis, for example, it implies that a catalyst, an expensive material, is deposited on appropriate support as nano-size clusters whose concentration is so little that they can be considered isolated. It turns out that, when the nano-particles do not agglomerate, such nano-systems are very stable. Then, how the macroscopic fluctuations take place? Hereafter we will outline that to the most surprise, such systems behave steadily and do not exhibit macroscopic fluctuations.

The kinetics of the reaction that proceeds on each nano-particle is very complicated because of the complex effects of the boundary, possible fractal dimension etc. However, we certainly know that in general it is described by equations of the reaction-diffusion type presented by Equation (7.14). Hence, the products that come out from each nano-particle permanently fluctuate. However, whenever the size of nano-particles exceeds the kinetic threshold of stability, each of them operates steadily and permanently in the course of time. In addition, the large distance between the nano-particles, makes them isolated from one another. So, the nano-particles contribute to the efficiency of the system independently. In turn, the sum of huge number of random numbers does not signal out any significant fluctuation. Thus, we come to surprising and highly non-trivial result: the kinetics whose outcome is permanent fluctuations brings about a steady non-fluctuating regime!

Let us now consider the important issue why the efficiency of the nano-particles is so sensitive to their size and to the external constraints. Indeed, it is well known that the nano-particles are efficient only when their size is confined to certain specific to the system range. The general condition for stable proceeding of a "reaction," viewed as a transformation of one species into another, is provided whenever the "kinetic" threshold of stability is smaller than the "spatial" size. However, the value of the kinetic threshold of stability is highly sensitive to the values of the external constraints such as temperature, pressure of the reactants in the gas phase etc. through their explicit involvement in Equation (7.14). Therefore, the ratio between the "spatial" and the "kinetic" size expressed as a function of the external constraints yields the range of efficiency of the nano-particles. Evidently, this range depends strongly on the values of the external constraints.

Thus we come to the most decisive for our approach result: this is that the requirement the spatial size of the nano-particles to exceed the kinetic threshold of stability imposes general constraint to further miniaturization of the nano-particles!

CONCLUSION

The major goal of the present and the next Chapters is the development of a new approach to the thermodynamics of the complex systems in a non-specified ever-changing environment subject to the mild condition of boundedness alone. The need for a new approach is imposed by the fact that the complex systems permanently exchange matter and energy with the environment which makes the major principles of traditional thermodynamics inappropriate. To remind, that the traditional thermodynamics is founded on the assumption that every system resides stably in a single state called equilibrium, which is such that a system arrives at it starting from any initial condition. Thus, in the traditional thermodynamics the thermodynamical potentials appear as state variables, i.e. they are independent from the path functions. This is possible if and only if the state variables are associated with certain extreme properties such as minimization or maximization of the corresponding thermodynamical potential. Although in our approach the thermodynamical potentials are also state variables, their changes are path dependent. Thus, the thermodynamical variables are not associated with any extreme properties of a system such as entropy maximization and/or potential extremization. The release from the condition of entropy maximization goes in accordance with the role of hierarchical organization of the complex systems. The latter appears explicitly through putting ban over "self-propelled" semantics, i.e. the hierarchical self-organization makes use of the property of stationary of the action so that not to allow any semantic-like response that commence from "noise" alone.

The great advantage of our approach is that the basic thermodynamic notions such as internal energy, temperature and the chemical potential have been derived in the setting that emerges from the properties of the state space alone. It is worth noting once again that the derivation does not involve any of the "traditional tools" like extreme properties, pairing intensive-extensive variable etc. It is grounded on the single requirement for stationarity of action. A decisive confirmation of the self-consistency of our approach is the result about the boundedness of the exchanged between a system and its environment matter/energy provided by the ban over the association/dissociation of arbitrary amount of particles by a system. Indeed, an exclusive property of the derived by us chemical potential is that it varies with associating/dissociating particle so that to sustain bounded variations of the matter/energy exchanged with the environment. The assumption about boundedness of the exchanged matter/energy is among the central assumptions of our theory. Therefore, its emergence as a result of this approach strongly supports the self-consistency of the developed by us theory.

REFERENCES

Feller, W. (1970). *An introduction to probability theory and its applications*. New-York, NY: John Wiley & Sons. doi:10.1063/1.3062516

Gardiner, C. W. (1985). *Handbook of stochastic methods for physics, chemistry and the natural science* (Haken, H., [Ed.] *Vol. 13*). Berlin, Germany: Springer Series in Synergetics. doi:10.2307/2531274

Chapter 9
Second Law Viewed as Ban over Perpetuum Mobile

ABSTRACT

It is proven that under boundedness, the efficiency of a non-mechanical engine never exceeds the efficiency of the corresponding Carnot engine where the engine is free from necessity of a physical coupling to two heat reservoirs. The proof is free from the condition for entropy maximization viewed as condition for reaching equilibrium. Thus the proof substantiates the most ubiquitous formulation of the Second Law to be ban over perpetuum mobile. Further, the ban over the information perpetuum mobile appears a consequence of the most general formulation of the Second Law which asserts that it is impossible to build a non-mechanical "engine," which works steadily in a cyclic regime without exerting any functional changes of its homeostasis during the working cycle.

INTRODUCTION

Nowadays the Second Law of the thermodynamics is considered as one of the fundamental laws of Nature since it is supposed to govern the arrow of time. It is assumed to reveal one of the principal early mysteries of thermodynamics, namely the origin of irreversibility: while microscopic equations of motion describe behaviors that are the same in both time directions, why do large-scale systems exhibit temporal asymmetries? Many thermodynamic processes go in one direction: closed systems devolve from order to disorder, heat flows from high temperature to low temperature, and shattered glass does not reassemble itself spontaneously. These are described as transient relaxation processes in which a system moves from one macroscopic state to another with high

DOI: 10.4018/978-1-4666-2202-9.ch009

probability since there is an overwhelming number of microscopic configurations that realize the eventual state. The Second Law asserts that the natural processes spontaneously evolve in the direction of increasing entropy so that eventually to reach equilibrium, the state of maximum entropy, where a system stays arbitrarily long time.

The success of the Second law is due to a great extent to its relation with the other key concepts of thermodynamics, the most important of which is the concept of thermodynamical potential. To remind, the notion of thermodynamical potential, considered along with the idea of thermodynamical equilibrium, gives rise to the ubiquitous Boltzmann-Gibbs measure whose role then stands as the probability for deviation from the equilibrium. Yet, the major advantage of the above setting is its enormous ubiquity: it allows each object in Nature to acquire thermodynamical properties such as heat capacity, susceptibility, etc. It is worth noting that these values are assigned to each object regardless to its internal structure and functionality. Thus, from the viewpoint of the traditional understanding of the Second Law, the structure and the functionality of an object are ignorant for its thermodynamical properties. As we already know, the complex systems exhibit long-range correlations encapsulated in the so-called power laws which in turn also are subject to create an arrow of time. Then, is it to be expected that the structure and the functionality are always ignorant for the thermodynamical behavior?! A question arises: are indeed thermodynamical properties such as heat capacity and susceptibility insensitive to the structure and functionality of the objects?

The most fundamental flaw of the Second Law is its controversy with the principle of relativity. By assuming that it sets the arrow of time, we actually select a point on the time scale where a process is initiated. Thus, every repetition of a process is also marked by enumerating its initiation point. This, however, violates the principle of relativity, which says that neither a process

nor any one of its repetition selects a special time point. This controversy has a practical implication: if the natural processes and their repetitions select time points, it would be impossible to define a power spectrum because, by definition, it selects correlations among distant responses but so that these correlations to be insensitive to the length and the position on the time arrow of the window at which the signal is recorded. Not only the sophisticated scientific tool such as the power spectrum, but our daily experience also suggests that finding solution of the controversy between these two fundamental laws of Nature: the Second law viewed as an implement for setting the arrow of time and the principle of relativity viewed as a ban over selecting of a special time points related to initiating or terminating a process, requires novel understanding of the thermodynamics and in particular requires novel understanding of the role of structure and functionality of the natural objects. The essence that could be taken out from these theoretical premises is to view the role of structure and functioning of the natural objects in a framework where both seemingly controversial facts peacefully coexist.

The goal of the present Chapter is to demonstrate that the above problem can successfully be resolved by the theory of boundedness. Not only it opens the door to re-formulation of the Second Law so that to eliminate it conflict with the principle of relativity, but it defines a specific measure for creating semantic units: the work done by the non-mechanical "engine" that constitutes the meaning of each semantic unit. Moreover, the operation of the corresponding non-mechanical "engine" is explicitly grounded on the functional properties of the corresponding system. At the same time, the developed approach "incorporates" the structure and functionality of a system in the traditional thermodynamical properties such as heat capacity and susceptibility.

The relation between the semantics and the arrow of time is grounded on the understanding of the sensitivity to permutations of the semantics. It is well known that each semantic unit is sensitive to permutations of its constituents: moreover, permutation sensitivity is an implement of distinguishing meaningful sequences of letters from the random ones. Therefore, the semantic meaning selects an "arrow" of meaning which is typical for each semantic unit. The problem is highly non-trivial since the "arrow" of meaning is a complex interplay between linear and circular behavior. While inside each semantic unit this arrow is linear, on accomplishing the unit, i.e. when the unit returns to the "space bar," the "arrow" turns circular. At this point, it is worth reminding that the semantic units on each hierarchical level of semantic organization are separated by punctuation marks: the words are separated by "space bars," the sentences are separated by dots, the paragraphs are separated by a blank line etc. Thus, it seems that semantics is in conflict with the Second Law since it does not select a monotonic arrow of time. This, however, poses the question whether it is ever possible to have a semantic-like response of natural systems? In this line of considerations it should be pointed out that a Turing machine also does not select a monotonic "arrow of meaning." Indeed, since it is *apriori* impossible to decide which algorithm halts, it is never known whether a given algorithm ends in "cycling" or accomplishes its task. Thus it is impossible to decide whether the algorithm selects a monotonic "arrow of meaning" or it runs an endless repetition of a "cycle."

Thus, we face the dilemma whether the information processes are an exception from the Second Law or the Second law calls for certain reformulation which is equally available for simple physical systems and for the complex systems exhibiting semantic-like response. Nowadays the scientific community is split in answering this dilemma. On the one hand, the Second law is considered as one of the fundamental laws of Nature and so the information systems, being natural objects, must obey it. On the other hand, the fact that the information systems do not select

a monotonic arrow of time turns decisive for some authors to exclude the information systems from the realm of the Second Law.

Turning point in our approach to the subject-matter is to break the view that arrow of time has a single origin associated with the notion of entropy. To make ourselves clear, we put forward the idea that the arrow of time has two origins:

1. The physical one which is related to the processes like devolving of closed systems from order to disorder, heat flowing from high temperature to low temperature, etc. We put forward the idea that the time asymmetry in these transient processes is implemented by the role of the non-unitary interactions introduced in Chapter 3 of the present book;

2. The "semantic" one which is related to the hierarchical order in complex systems. Indeed, the hierarchy sets specific relations between different cycles substantiated through intra-and inter-level feedbacks. The stability of this coordination is guaranteed by the stationarity of the action considered in Chapter 7. The major role of keeping the action stationary is that so it does not allow "self-propelling" cycles and thus keeps the complex motion of the hierarchically coordinated "semantic-like" cycles steady and stable.

The crucial point for the successful separation of the onset of the time asymmetry is the renounce of the traditional idea for associating the arrow of time with the motion towards the state of maximum entropy. This step is necessary because the complex systems are highly structured systems whose functioning is far from being in a state of a single equilibrium. To remind, they can reside in states of different "homeostasis" instead to reach a single equilibrium state. Thus, since the homeostasis is a self-organized state which cannot be described by a single parameter, e.g. probability, it is impossible to fit its properties in the frame of the traditional statistical mechanics which enumerates the points in the state space by assigning a single parameter, probability, to each of them. On the contrary, the state space of complex systems has locally Euclidian metrics, as established in Chapter 4, so every point is characterized not only by certain, specific for the state parameter, but with the distances to the nearest neighbors as well. Therefore, the entropy stands as a multi-valued function which has specific value at every point and whose selections are dictated by the admissible transitions. Thus, the entropy acquires the property of being defined in a non-extensive and non-probabilistic manner.

The non-probabilistic way of defining the notion of entropy is advantageous not only for explanation of the problematic with the arrow of time, but it allows to extend the ban over perpetuum mobile beyond simple mechanical devices, to complex systems capable to semantic-like response. We will demonstrate that the ban over perpetuum mobile is spread over non-mechanical "engines" which are not necessarily coupled to two heat reservoirs. In turn, this provides the enormous ubiquity of this ban, and this serves as grounds for proclaiming the ubiquitous ban over perpetuum mobile as the adequate formulation of the Second Law as one of the fundamental laws of Nature.

BACKGROUND

One of the central assertions of our theory is that the response is local and active. It implies that for small environmental impacts the corresponding response is proportional to the impact and that the response holds only when the impact is applied. Even though this assertion is very similar to the traditional notion of fluctuation-dissipation theorem, it is still not equivalent to it: whereas local fluctuations are considered random independent from one another events in the traditional thermodynamics, the boundedness introduces non-

physical correlations among distant responses. Next we shall demonstrate that the traditional approach is not able to substantiate the above assertion. The importance of this consideration is that the obtained flaw of the traditional approach violates the very core of the traditional thermodynamics, namely that of the necessity of extremization of the entropy viewed as a necessary condition for obtaining the normal distribution of the fluctuations.

To remind, a central assumption of the traditional thermodynamics is that the local fluctuations appear as random independent from one another events and thus their distribution, being a subject of the Central Limit Theorem, is normal. The condition for considering the local fluctuations as random independent from one another events is the extremization of the entropy. It is well known fact that the thermodynamical variables in the traditional approach are separated into two classes: intensive and extensive with respect to the volume of the system, so that they to appear in pairs intensive-extensive variables in the corresponding thermodynamical expressions which are supposed to be homogeneous functions of the system's size. This is the way the traditional approach meets the thermodynamical limit. What is usually not pointed out explicitly, is the relation of this assumption with the matter of extremization of the entropy. Let us now, elucidate this relation in details: according to the above framework, the entropy is an extensive variable and at the same time in the state of equilibrium it reaches its maximum. To meet this task, it is necessary that the entropy reaches its maximum in any sub-cell obtained by arbitrary partitioning of the system into sub-cells. Further, the maximization of the entropy in each sub-cell requires that the energy of that sub-cell is well defined and does not change significantly for the time period in which the entropy maximization happens. Put in other words, this condition implies that the sub-cells should be considered closed systems. This suggests that the corresponding fluctuations are random events independent from one another. Then, the distribution of the fluctuations turns out to be normal.

However, the normal distribution of the fluctuations is in conflict with the fluctuation-dissipation theorem. The latter asserts that the response of a system to a small environmental impact is equivalent to the development of corresponding fluctuation. This implies that the probability for a fluctuation, viewed as a characteristic of the internal "driving forces" for creating of a given fluctuation, must be linear function with respect to the size of the fluctuation in order to provide the matching between the environmental impact and the internal "driving forces." However, the Taylor expansion of the normal distribution for small fluctuations is essentially non-linear:

$$P\left(x\right) \propto \exp\left(-\frac{\left(x - x_{eq}\right)^2}{\sigma^2}\right)$$
$$P\left(x\right) \propto 1 - \frac{\left(x - x_{eq}\right)^2}{\sigma^2} \tag{9.1}$$

The problem does not reside only in the formal inconsistency of the Taylor expansion for the small fluctuations, but it has deeper origin. The inconsistency must be traced back to the idea of partitioning of a system into independent from one other sub-systems. This is possible only if there are no specific temporal and spatial scales in the system created by the processes that proceed inside. However, the systems that meet this requirement are only ideal gases. The vast majority of systems, in particular complex systems, are highly structured and their parts "interact" with each other so that to produce the coordinated behavior of the corresponding system. Starting with the simplest reaction-diffusion systems there is at least one pair of spatio-temporal scales set by the balance between the diffusion and the reaction. Then,

the balance between the surface phenomenon viewed as diffusion and the volume phenomenon viewed as a reaction, selects a specific spatial and time scale where it occurs. Since the selection of those specific scales is a generic property of every reaction-diffusion system, it discards the relevance of the view over the thermodynamical limit as invariance of the macroscopic variables under partitioning.

Then, obviously, we must abandon the traditional approach to the problem. Instead, what comes foremost, is establishing the general condition for the stable behavior of a complex system. We already introduced this condition and it is the boundedness of the local fluctuations and their covariance. To remind, the latter implies that the occurrence of any local fluctuation does not signal out any special time and space point. This is necessary for meeting the condition that the stable behavior of a complex system obeys the principle of relativity, i.e. the stable behavior of any system is time-translational invariant.

In Chapter 3 we have revealed the high non-triviality of introducing the concept of boundedness for local fluctuations. As we have established there, the simple idea of constraining the velocities of the constituents alone yields amplification of the local fluctuations instead of their boundedness. That is why, the relevant approach to the problem of boundedness involves considering the concept of boundedness at the quantum level. In result, we have found out that the local fluctuations are bounded and covariant. Then, their succession in the time course could be presented as a bounded irregular sequence (BIS). The properties of BIS are studied in Chapter 1. It has been found out that deviation from the average of any BIS over a window of length T is $O\left(\frac{1}{T}\right)$ while the average never exceeds $O\left(T\right)$. Then, the average maximal local fluctuation is of the order of a constant:

$$P \propto O(T) O\left(\frac{1}{T}\right) \approx \lambda \qquad (9.2)$$

It is obvious that the parameter λ is a constant which does not depend on the position and the size of the window. Then, we formally could assign Poisson distribution with the parameter λ to the local fluctuations. Note, that the origin of the Poisson distribution is the boundedness and the covariance of the local fluctuations. Thus, this justifies the fundamental difference between the present approach to the Poisson distribution and its traditional derivation from the binomial distribution. The great advantage of the presently obtained Poisson distribution for the local fluctuations lies in its linear asymptotic for small fluctuations:

$$P \propto \exp(-\lambda x)$$
$$P_{small}(x) \propto 1 - \lambda x \qquad (9.3)$$

Therefore, it is possible to conclude that our approach justifies the linear response theory for small fluctuations. Moreover, the present considerations provide credible basis for its ubiquity and relevance for wide variety of systems.

The importance of validity of the linear response theory is that it extends the notion of a closed system beyond the constraint of constancy of some characteristic, e.g. energy, and the subsequent need for extremization of other characteristics, e.g. entropy for providing the homogeneity of the thermodynamical functions with respect to the system's size. Instead, the linearity of the response is achieved by means of providing the boundedness and the covariance of the local response and the local fluctuations. It is worth noting that the boundedness and the covariance of both the response and the fluctuations is achieved at the presence of specific spatial and temporal scales in the system imposed by the specific for the system processes and its structural and functional hierarchical organization. In turn, this releases the necessity for considering the thermodynamical variables in pairs intensive-extensive ones so that to meet the thermodynamical limit. Instead, we are free to define the thermodynamical variables neither as intensive nor as extensive ones. Let us remind that the origin of any of them is the equation-of-state which comprises the structural and the functional properties of the system. In Chapter 7 we have already encountered one of them: the energy. To remind, we define it by means of Equation (7.5) whose terms come from the equation-of-state. Further, the internal energy defined through Equation (8.11) also comes from the equation-of-state. Therefore, the basic thermodynamical variables explicitly encapsulate the structural and the functional properties of the system involved in the equation of state.

This provides the major difference with the traditional statistical mechanics where the basic thermodynamical variables are identified with certain microscopic characteristics. However, these relations have an inevitable artificial background since it is assumed that the dynamics does not contribute straightforwardly to the macroscopic behavior. It is supposed that because of its complexity, the evolution is governed by parameters that have no dynamical analog - entropy, concentration, etc. whose introduction requires probabilistic manner of describing. The conclusion is that there is no explicit relation between the probabilistic and the dynamical level of description. Instead, our approach replaces the probabilistic level of description by means of relating the basic thermodynamical variables and relations explicitly with the structural and functional properties of a system through the equation-of-state.

ARROW OF TIME

The traditional thermodynamics relates the arrow of time with the origin of the physical irreversibility. The latter implies that many thermodynamic processes go in one direction: Closed systems devolve from order to disorder, heat flows from high temperature to low temperature, and shattered glass does not reassemble itself spontaneously. This irreversibility is associated with the major paradigm of the many-body science: the macroscopic behavior of the many-body systems is described through characteristics such as entropy, concentration etc. which has no dynamical analog and require probabilistic manner of description. Then, the thermodynamic irreversibility is a result of the monotonic increase of one of these variables, namely the entropy.

Yet, since its early time this paradigm has been in a fundamental conflict with the underlying dynamics. The scientific community shares the view that the underlying dynamics of the constituents is reducible to the Newtonian one in the case of the classical systems and to the conventional quantum mechanics for quantum systems. A crucial for our consideration property of both the classical and the quantum equations of motion is its reversibility with the respect to time. In turn, the time-reversibility of the dynamical motion drives a fundamental conflict with the idea of the monotonic increase of the entropy until reaching maximum because the entropy, by definition, is related with the degree of disorder of the system. Then, the maximum entropy corresponds to the state of prefect mixture. However, the monotone reach of the state of perfect mixture and the subsequent self-sustaining of that state requires apparent irreversibility of the dynamics. On the other hand, however, the dynamical reversibility puts a fundamental ban on the monotonic approaching and self-sustaining of the state of maximum entropy.

Another fundamental flaw of this setting is the conflict with the principle of relativity. The idea of association of the arrow of time with the monotonic approach to the state of maximum entropy implies that the initiation of the process selects a special point on the time scale. Thus, at each repetition it also selects a special time point. Then, the successive repetitions of the process could be enumerated. The latter, however, violates

the principle of relativity which asserts that the laws of Nature are time-translational invariant.

In order to eliminate the above difficulties we put forward the idea that the arrow of time has two different origins: the first one is related to the transient processes such as devolving from order to disorder, heat flowing from high temperature to low temperature etc.; the second one is related to the role of the hierarchical order in the complex systems. It should be pointed out explicitly that we consider the establishing of the arrow of time at the expense of releasing it from any association with the properties of the entropy.

Arrow of Time for Transient Processes

Now we start with establishing the arrow of time for the transient processes such as dissolving of a drop of ink in a glass of water. Our daily experience teaches us that the spontaneous evolution of the process is unidirectional – it always goes in the direction of formation of perfect mixture between the ink and the water. The intuitive idea is that origin of this unidirectional evolution should be traced into dynamics. However, the dynamics is supposed to be time-reversible which immediately puts a ban over the monotonicity of the process. But is the dynamics of the many-body systems indeed time-reversible?! The strict formal answer is affirmative only for an isolated act of collision. But are the collisions in the many-body systems indeed strictly isolated? The affirmative answer has been adopted for the ideal gases which are so rare that the two-body collisions are supposed well isolated from one another for the time of collision. In the recent decades, the theory of many-body interactions gains a serious boost by developing a number of sophisticated methods such as the diagrammatic approach. The major common property of the many-body interactions remains their unitarity. In turn, this assertion allows additive decomposition of each many-body interaction to a succession of "clusters" of $n-$

body interactions. Then, the unitarity of the evolution renders each and every interaction independent from one another. However, not only the unitary evolution does not provide the monotonic evolution of our solution but it opens the door to arbitrary large local accumulations of matter and energy at any locality and at any time; it is asserted that this effect is negligible since it is assumed that these accumulations are automatically averaged out. However, an overlooked point is that since all local fluctuations are independent from one another, prior to averaging out, any large enough local accumulation of matter and energy can trigger novel, even hazardous for the stability events: phase transitions, pattern formation, sintering, overheating etc. whose major property is their unpredictability and non-reproducibility both on micro and macro-level. In result the system would behave in an uncontrolled, unstable, unpredictable and non-reproducible way. Thus the demand for a stable, controlled and reproducible evolution sets the requisition about a self-sustained mechanism for permanent control over local accumulation of matter and energy.

In Chapter 3 we have proved that the assertion about the independence of the interactions is only approximation that is overruled even for very low concentrations: indeed, along with the unitary processes there always are non-unitary processes which come out as a result of the inevitable involvement of the random motion of constituents (atoms, molecules). Indeed, any collision could be interrupted by another constituent that enters the interaction as a result of its random motion. The non-unitarity is due to the time-reversal *asymmetry* between the ingoing and outgoing trajectories of the colliding constituents which commences from the dependence of the outgoing trajectory on the status of interaction at the moment of arrival of the new constituent. To make this clearer, let us consider a 3-body classical interaction, subject to energy and momentum conservation law. A simple check shows that in this case the laws of energy and momentum con-

servation do not unambiguously determine the outgoing energies and velocities. Indeed, the classical energy and momentum conservation for a $3 - \text{body}$ interaction read:

$$\frac{m_1 \vec{V}_1^2}{2} + \frac{m_2 \vec{V}_2^2}{2} + \frac{m_3 \vec{V}_3^2}{2} = \frac{m_1 \tilde{\vec{V}}_1^2}{2} + \frac{m_2 \tilde{\vec{V}}_2^2}{2} + \frac{m_3 \tilde{\vec{V}}_3^2}{2}$$
(9.4)

$$m_1 \vec{V}_1 + m_2 \vec{V}_2 + m_3 \vec{V}_3 = m_1 \tilde{\vec{V}}_1 + m_2 \tilde{\vec{V}}_2 + m_3 \tilde{\vec{V}}_3$$
(9.5)

where m_i, $i = 1, 2, 3$, are the masses of the entities; $\vec{V}_i$ are their initial velocities and $\tilde{\vec{V}}_i$ are the final velocities. However, we have 4 equations for the 9 components of the final velocities. Hence indeed the energy and momentum conservation are not enough to determine non-ambiguously the final velocities even in the simplest case of an elastic $3 - \text{body}$ interaction.

The good of the above non-ambiguity in determining the exact velocities and energies of each of the colliding constituents, is that the many-body interactions act as "randomizers" of the energies and velocities. And what is more important, it proves that the many-body interactions are time-irreversible. Note, that the two-body interactions are time-reversible in the same setting, i.e. under the constraint of energy and momentum conservation. Then, the fundamental time-irreversibility of the many-body interactions is the first crucial ingredient for the natural choice of the direction of the transient processes.

The time asymmetry of the many-body interactions alone is not enough to provide the monotonic course of the transient processes and their time-translational invariance. These properties require covariance of the many-body interactions. The latter is provided by the coupling between the non-unitary interactions and the acoustic phonons which ensures the covariance of the corresponding dissipation viewed as insensitivity to a particular choice of the reference frame. As discussed in Chapter 3, this is exclusive property of the boundedness, implemented by the coarse-graining, which renders the current state of any atom (molecule) independent from the interaction path. This property verifies our assumption that the non-unitary interactions are metric free in the sense that their participation in the corresponding Hamiltonian is not specified by metric properties such as distances, velocities etc. Yet, the apparent dependence on the number of surrounding constituents prompts the suggestion that it is controlled by the current local concentration. Thus, the inter-level feedback becomes evident: at the quantum level it controls local fluctuations so that each of them stays bounded at any moment; thus the concentration is kept permanently well defined; in turn, the well defined concentration sustains and controls the intensity of the non-unitary interactions and thus puts the local accumulation of matter/energy subject to boundedness.

Outlining, we conclude that we have the necessary and sufficient conditions for providing a time-translational invariant monotonic unidirectional evolution of the transient macroscopic systems such as dissolving of ink into water. The necessary ingredient is the non-unitarity of the many-body interactions. The sufficient condition is provided by the feedback between the local gapless modes and the dissipation of the non-unitary interactions whose covariance guarantees the time-translational invariance of the process. Put in other words, we have proved that our theory provides accordance of the unidirectional evolution of the transient physical processes and the principle of relativity.

Arrow of Time for Semantic-Like Response

The reason for separating origin of time arrow for transient physical phenomena and the complex systems capable to semantic-like response is that while a transient phenomenon has its initial and final state so that eventually the corresponding

system rests at the final state for arbitrary long time, the complex systems exert permanent motion among states of different algorithmic value, i.e. they go through different basins-of-attraction. Then, obviously our task is to evaluate both the asymptotic properties of that motion and its local ones. The importance of establishing the asymptotic properties gives the answer to the question whether it is ever possible to gain an arbitrary large amount of knowledge. The latter is of decisive importance not only for academic reasons but also for creating a long-term strategy for providing sustainable evolution in an ever-changing environment.

Let us start with finding out the role of the hierarchical organization for setting a local arrow of time. It is obvious that a local arrow of time always exists due to the elaborated by us idea of associating the meaning of a semantic unit with the specific performance of a non-mechanical "engine." Then, the major property of any engine to operate specifically according to the direction of its running substantiates the idea about a local arrow of time.

Before establishing the asymptotic properties of a semantic-like response let us remind the major necessary conditions for a complex system to exert a semantic-like response. The clue is that self-organized semantics is possible only when the state space has certain minimal structure and topology. And the central point is that it must comprise at least 3 basins-of-attraction and an accumulation point. The reason for this is as follows: fundamental ingredient of any semantics is the space bar, which serves as a separator among words. What could be the "space bar" in the state space? It must be a state adjacent to each "letter" so that to provide the freedom to a word in order to start and end with (almost) every "letter." An immediate outcome of this view is that the motion in the state space is to be orbital. Thus both elements are necessary for the simplest semantics conditions impose certain structure of the state space: it must consist of at least 3 basins-of-attraction and to have accumulation point which is tangent to each of the basins. Then, the motion is indeed orbital so that each orbit is tangent to the accumulation point. What provides the crucial difference with a random sequence of letters is that the bounded motion allows transitions only to adjacent basins. Thus, the choice of further letters is limited to those whose basins are adjacent.

The spontaneous limitation of the admissible neighbors for every letter is the first level of setting grammatical type rules in the organization of the response. There is a very interesting parallel between the grammatical rules, causality and the 'arrow of time'. Each time when a system works as "oracle" the corresponding transition from one basin-of-attraction to the only admissible one could be considered as a deterministic causal relation between the corresponding laws. An exclusive property of that deterministic relation is that it holds even in varying environment. In cases where the admissible basins are more than one, the random choice of one selection could be considered as "free will." Let me remind once again that the properties of being "oracle" and of the "free will" are properties of the self-organization of certain natural systems and thereby are not to be taken as exclusive properties of some superior human mind.

The admissible orbits that substantiate words reveal long-range correlations among the letters in the corresponding words; we have already considered that these trajectories are non-Markovian. This opens the door to assigning "arrow of time" to each orbit. Its origin is to be associated with the sensitivity to permutations in the succession of letters. Yet, there is highly non-trivial interplay between the circularity and the linearity of the "arrow of time": it is locally linear, i.e. it is linear while writing a word; when finishing writing it turns circular; at the next level (writing a sentence) it is again linear, when putting a dot it becomes circular. It is worth noting that such interplay between circularity and linearity of the arrow of time is substantiated by the idea of associating the semantics with the natural processes. It should be stressed that neither philosophical tradition, not western nor eastern, assigns the property of time to be simultaneously linear and circular. In the context of the modern interdisciplinary science we could say that the time is not scale-invariant. It is more appropriate to say that time is a characteristic of the dynamics of the response and thus its properties depend on the local dynamics. Viewed from this prospective, time has the same origin and role as the metrics in the state space: they both commence from the dynamics and are function of the hierarchical organization. The major role for the interplay between the circularity and linearity of the time is the non-extensivity of the hierarchical order.

Thus, the above structure of the state space provides two major differences with a Turing machine: the first one is that a hierarchically self-organized system spontaneously "says words" in response to the environment. Further, these words are not random sequences of letters, space bar included. What makes the sequence to have autonomous semantics is the restraint established over the possible succeeding of letters: for example "a" cannot follow "o." Thereby, the set of possible words is not the total set of all possible combinations of letters. Thus, the major difference between the so organized primitive semantics and a Turing machine is that "hardware" becomes active: it selects a subset of semantic units (words in the case considered) so that to "express" itself. The advantage of this form of "response" is that the system stays stable. Moreover if a letter is missing in recording we can easily identify it because we know the admissible set of its predecessors and successors; we can even "foresee" the future state of the system on the grounds of the knowledge about the current state only. Thus, the property of being an "oracle" substantiates the major advantage of the semantic-like response: it provides its non-extensivity. The non-extensivity constitutes the major difference with the traditional computing. Indeed, the extensivity of the computing commences from the assertion that every algorithm is reducible to a number of arithmetic operations, executed by means of linear processes. Thus, the more complex is the algorithm the more operations it involves and in turn, the more hardware elements (and/or time) are necessary to accomplish the task. Along with the extensivity, the traditional computing is non-autonomous: it execution is governed by an "external mind" who sets the algorithm and who comprehends the results.

Now we can delineate the differences between setting arrow of time for a Turing machine and the semantic-like response. The first major difference is that while the algorithm of a Turing machine is linear and non-restrained in length, the non-extensive cyclic –like organization of the functioning of the complex systems implies that existence of the largest cycle of finite length. One can conclude that she/he can acquire infinite knowledge by means of Turing machine. Yet, this is always subject to uncertainty since it is never known *a priori* whether a given algorithm halts or not. On the other hand, the algorithmic value of the semantic-like response asymptotically tends to a constant value. This surprising result comes out from the power spectrum of BIS – whatever the succession of information symbols is given, the power spectrum of an infinitely large BIS tend to a continuous band of shape $1/f^{\alpha(f)}$. Then, one can conclude that it is impossible to acquire an infinite knowledge. The comfort of this surprising statement is that the ban over acquiring of an infinite knowledge is put at the expense of sustaining balance between the corresponding complex system and its environment. This asymptotic is locally non-uniform: this means that for any time window of finite length we can acquire certain knowledge. Therefore, we are now able to explain why a Beethoven symphony, a product of a genius mind, and the traffic noise have the same power spectrum. Alongside, now we are able to explain why very long pieces of information: films, books, conversations, become boring – their algorithmic

content reaches saturation on increasing their algorithmic duration.

Thus, the idea about the relation between the arrow of time and the algorithmic value of the corresponding "semantic meaning" turns able to explain not only the non-uniformity of the arrow of time but the emotional status of comprehending the semantics as well.

ENTROPY UNDER BOUNDEDNESS

The notion of entropy is the pinnacle of the traditional statistical theory. Its value is substantiated by its role as a bridge between the microscopic and the macroscopic description of the behavior of the many-body systems. It is grounded on the central assumption that the macroscopic behavior of any many-body system cannot be described by the dynamical properties of its constituents alone; instead the macroscopic description requires the use of variable neither of which has dynamical analog – and one of these variables is the entropy. This viewpoint opens the question whether it is possible to associate the notion of the entropy with the functional organization of the system. The key assumption for such association is the role of the requirement that the entropy must reach its maximum at equilibrium and that it must be an extensive variable, i.e. its value must be proportional to the size of a system. The simultaneous fulfillment of these requirements is possible if and only if the entropy of a system is insensitive to arbitrary partitioning into independent from one another sub-systems. The latter, however, implies that there exists an invariant to partitioning, that could be assigned to each sub-system, so that the entropy to be maximized at the constraint of constancy of that invariant. Usually, this invariant is supposed to be the energy; its choice is suggested by the notion of a closed system whose major characteristic is namely the constancy of the total energy. Thus, by analogy, it is supposed that the energy of each sub-system is invariant under partitioning. Yet, the plausibility of such consideration is questioned by the fact that it is fulfilled only if the processes developed in a system do not substantiate existence of specific spatial and temporal scales in the system which would compromise the invariance to arbitrary partitioning. Thus, obviously, the invariance under partitioning turns a rather exclusive property of the ideal gases and the well stirred solutions than a typical property of the complex systems. This fact, however, puts under questioning the validity of the entire concept about the probabilistic description of the entropy.

Another flaw of the traditional notion of the entropy is even more fundamental and is related to the role of the fundamental uncertainty of defining the accuracy of the energy. To make this assertion clear, let us consider the traditional derivation of the notion of entropy: the entropy is defined as the logarithm of the number of the admissible states. Thus, the entropy is defined as an intensive variable – the entropy of two independent systems is the sum of their entropies. The grounding assumption for this definition is that the admissible microstates are equiprobable. However, this derivation of the property of the entropy to be an extensive variable suffers an often overlooked fault: It is related to the principal uncertainty of the accuracy by which the energy of a closed system is defined. To remind, there exists at least one fundamental origin of uncertainty for defining energy: the Heisenberg relations $\delta U \delta t \approx \hbar$. Let us now consider the impact of that uncertainty on the definition of the entropy. The traditional approach to the matter is to assume that the number of the admissible states in the interval δU is $dU = \Delta U \delta U$ where ΔU is the average number of admissible states in the interval δU around U. Then, the entropy is:

$$S(U) = \ln dU = \ln \Delta U + \ln \delta U \qquad (9.6)$$

Usually, it is taken for granted, that because for macroscopic systems the difference between the magnitude of the first and the second term in Equation (9.6) is of many orders of magnitude, the energy uncertainty can be ignored. However, the ignorance of the second term is practically non-negligible for nano-systems and thus fundamentally non-negligible for macroscopic systems when the insensitivity to arbitrary portioning is required. Actually, both cases are related: the arbitrary partitioning implies that the sub-systems could be considered nano-sized ones as well. The non-overwhelming difficulty with the energy uncertainty is that, no matter how small it is in a certain moment, in a long run it is accumulated and thus compromises the monotonicity of the entropy viewed as the spontaneous choice of the direction for the evolution of a closed system.

Thus, the theory of the macroscopic description of the complex systems faces the fundamental need for reconsideration. The first question in this line of reasoning is whether the macroscopic viewpoint must be kept a probabilistic one. Our answer to this question is firmly negative: we assert that the first step to a successful theory is the withdrawal from the probabilistic view of the entropy. The reason for that is two-fold: on the one hand, any probabilistic description implies that the state space has no metrics. In turn, the only alternative for the transitions between states would be their Markovianity. This, however, implies that no hierarchical organization is ever possible since the Markovianity allows arbitrary succession of local transitions which could result in accumulation of extra energy/matter in a single locality. The avoidance of the accumulation of extra- energy/matter is the starting point for creation of our approach of boundedness whose development is the subject of the present book.

Another flaw of any probabilistic approach to the notion of entropy lies in its relation to the linear-response theory. Indeed, if the small fluctuations are defined through any probabilistic distribution, the validity of the fluctuation-dissipation theorem implies that the response and impact are pre-determined by this distribution. This, however, tacitly presupposes that the impact is globally pre-determined and thus lacks identification. On the contrary, the central assumption of the theory of boundedness is that the impact and the response are local and are defined only by the current impact and the current state of the system.

The inconsistency between the Markovianity and the boundedness is provided by the central assumptions of the proposed by us theory of boundedness. The hypothesis of boundedness consists of 1) a mild assumption of boundedness on the local (spatial and temporal) accumulation of matter/energy at any level of matter organization and 2) boundedness of the rate of exchange of such an accumulation with the environment. Under the assumption of the hypothesis of boundedness the state space of an open system is partitioned into basins-of-attraction and the trajectories form a dense transitive set of orbits. The second part of the hypothesis is implemented by restricting transitions from a point on a trajectory to being those to nearest neighbors only; i.e., transitions from any point on a trajectory must pass through an adjacent state (one can think of a latticised state space). For adjacent states to be well defined, closed trajectories must belong to the same homotopy class which restricts the loci of trajectories to being simply connected. To compare, the Markovian trajectories are inherently topologically unconstrained and can therefore have as loci manifolds that are not simply connected (e.g., a torus) which in turn means that trajectories can belong to different homotopy classes. Further, the extremization of the probability for residing in the equilibrium requires holding of the detailed balance. Alongside, the condition of detailed balance appears as a condition for elimination of any steady fluxes. Thus, a Markovian trajectory never closes. This substantiates the major difference with the motion in a state space under bounded-

ness which is orbital on coarse grained scale and random on the finer scales. In turn, this difference sets the exclusive property of the semantics-like response – the semantic meaning to be separated from the structure of the semantic unit. Indeed, the theory of boundedness allows considering the meaning of a semantic unit with the performance of an engine which exerts a cyclic process and which thus is algorithmically irreducible to its algorithmic structure presented by a given sequence of "letters." This two-fold presentation of the semantic organization lays the grounds for the non-extensivity of that organization.

A question arises: whether it is ever possible to gain an absolute knowledge, i.e. whether the "engines" behind the semantic units work as perpetuum mobiles and whether it is ever possible to transform the "noise" into semantics, i.e. whether it is ever possible to acquire knowledge from the noise? In the remaining part of this chapter we will consider this problem and its relation to the notion of the entropy.

The first problem we encounter in our efforts to introduce the notion of entropy in the theory of boundedness is related to the existence of metrics in the state space and the constraint for setting the admissible transitions to the nearest neighbors only. Since the constraint of the admissible transitions to the nearest neighbors compromises the idea of state variable viewed as a variable whose change is path independent, we must reconcile the idea of entropy as a state variable whose changes are path dependent. The immediate consequence of such viewpoint is that the path dependence eliminates the idea about the existence of equilibrium as a global attractor where the entropy reaches its maximum. In turn, this implies the withdrawal from the probabilistic view over the notion of the entropy. The formal mathematical reconciliation of the idea about a path dependent state variable is implemented by the so-called multi-valued functions: these are functions whose value is defined in each point and whose changes are defined as random choice among given selections associated with that point. The constraint imposed by the boundedness renders all selections bounded. The implementation of the state variables through multi-valued functions is an exclusive property of the boundedness.

Now we are ready to define the notion of entropy. We define it through the first law of thermodynamics given both for the state variables and their local changes. Thus,

$$U = TS + \mu N$$
$$\Delta U = S\Delta T + T\Delta S + \mu\Delta N + N\Delta\mu \quad (9.7)$$

where U is the internal energy, μ is the chemical potential.

A crucial for our approach property of the second equation in (9.7) is that each of the changes of the corresponding state variables is governed by the equation-of-state which dictates the motion in the state space. The importance of that property is that the equation-of-state serves as grounds for setting the metrics in a state space under boundedness as considered in Chapter 4. Its exclusive property is that the establishing of metrics is grounded on the constraint of the admissible transitions to the nearest neighbors only. Thus, the presence of metrics in the state space helps to define not only the current values of the state variables but their admissible changes as well.

Another crucial for the role of hierarchical organization property of the second of Equation (9.7) is that the admissible changes of the state variables are coherent with the stationarity of the action introduced in Chapter 7. Note, that the stationarity of the action is the general implement of the hierarchical organization for sustaining coherent behavior of the cyclic-like hierarchical motion organized in a non-extensive manner. Thus, the present formulation of the first law of the thermodynamics expresses the energy conservation law whose admissible dynamics is subject to coordination with the stability of the corresponding hierarchical organization of a complex system.

It should be stressed that the values of all other state variables and their admissible changes are defined through the metric properties of the state space. Indeed, the value of the internal energy, U, is defined through the condition (8.11) which puts a ban over self-propelled steady fluxes whose origin is the noise alone. And since the motion in the state space is dictated by the equation-of-state, the admissible transitions are determined through it. Further, the notion of "temperature" also follows from the equation-of-state: to remind, it is defined as the variance of the power spectrum according to the considerations after Equation (8.17). The chemical potential, μ, is also defined through the properties of the state space as proved in the corresponding section of Chapter 8. A very important property of the chemical potential is its boundedness, i.e. it is impossible to associate/dissociate arbitrary number of species to a given system. This fact, along with the boundedness of the internal energy and any of its change, and along with the apparent boundedness of the effective "temperature," render the entropy and any of its change bounded. In turn, the boundedness of the entropy and its changes makes possible to associate the notion of entropy with the inevitable internal structural and functional reorganization of a system under exerting a non-mechanical work. At this point it is important to stress that the present definition of entropy is much wider than the traditional one since, while the traditional one considers the notion of entropy insensitive to any changes in the functionality of a system, the present one is explicitly grounded on the operational properties of the system expressed through the equation-of-state.

The enormous advantage of the present reformulation of the thermodynamics is that the present approach sustains the traditional meaning of the Maxwell relations. It is worth noting that the general condition for the justification of the Maxwell relations requires presence of a thermodynamical potential. The presentation of the major state variables through multi-valued functions successfully reconciles the existence of thermodynamical potentials and the path dependence of their changes. Thus, the traditional thermodynamical characteristics such as heat capacity and susceptibility are obtainable through the corresponding Maxwell relations. It should be stressed that although the Maxwell relations retain their traditional form, their exclusive property under boundedness is that they are sensitive to the functional properties of a system unlike the traditional approach where these relations are insensitive to the functional properties.

BAN OVER PERPETUUM MOBILE

In the traditional thermodynamics the assertion of banning perpetuum mobile is considered as one of the formulations of the Second Law. It states that it is impossible to gain work at the expense of extracting heat from a single reservoir alone. This assertion is equivalent to the more wide spread statistical understanding of the Second Law, namely the law that considers the entropy maximization as providing the arrow of time, but only for reversible processes. That is why there always have been attempts to "construct" a perpetuum mobile. The idea seems so attractive, that such attempts are never stopped. The question about the spreading of the ban over perpetuum mobile is of extreme importance for the information theory because of its explicit relation to the question whether there is an absolute knowledge and if so it is achievable. Put in other words, at stake is the question about the energetic and entropy "price" for getting information.

Before considering the ban over perpetuum mobile for systems able to exert non-mechanical work, it is worth noting that the traditional science is not free from perpetuum mobile. Perhaps the most exciting example comes from the theory of

relativity. The famous Einstein relation between mass and energy allows the mass to grow unrestrictedly when the velocity approaches the speed of light. The famous relations $E = mc^2$ where $m = m_0 / \sqrt{1 - v^2 / c^2}$ allows unrestrained growth of both energy and mass when the velocity v approaches the speed of light c. However, since this relation is supposed universal, i.e. available for every system regardless to its particularities and since infinite energy means infinite work for its achievement, does the acceleration process which substantiates this relation implies involving of a special kind of "perpetuum mobile"? It should be pointed out that the developed in the present book theory of boundedness puts a ban over accumulation of arbitrary energy/matter so that to allow transformations of only bounded amounts of energy and mass. Next we will prove that.

Generalization of the Carnot Efficiency

The developed in the last 3 chapters of the book generalization of the thermodynamics poses the question about the corresponding generalization of its traditional implements. A central issue of this matter is the idea of banning perpetuum mobile. Along with an academic curiosity, the interest in this topic is provoked by its enormous importance for setting a standard for the best practical efficiency of any newly tailored device.

Next we prove that the idea about the Carnot engine admits very powerful generalization in the following sense: *the efficiency of any engine that operates reversibly never exceeds the efficiency of corresponding Carnot heat engine where the engine is free from necessity of a physical coupling to two heat reservoirs!*

If exists, an effective Carnot heat engine must be derived under weakest possible constraints. That is why, we start the derivation on the grounds of the first law of the thermodynamics since, being the energy conservation law, it is universally available:

$$\Delta U = S\Delta T + T\Delta S + \mu\Delta N + N\Delta\mu + F_i\Delta\Lambda_i \tag{9.8}$$

where T is the temperature, S is the entropy; $\vec{F}$ are the thermodynamical forces and $\vec{\Lambda}$ are their conjugates whose values are set by the control parameters.

The only constraint imposed over Equation (9.8) is that of boundedness of each difference. This constraint commences from the natural requirement that the system stays stable if and only if the rates of exchange energy/matter/information with the environment are bounded.

Let us now consider that the system exerts a reversible cycle in the control parameter space. The crucial point is that the boundedness of all derivatives renders correspondence between the projections of the cycle in the (F_i,Λ_i) section and (T,S) sections so that to ban any degeneration of a cycle in (F_i,Λ_i) section into a line in (T,S) section. The reason is obvious: any line in the section (T,S) implies infinity of the derivatives which, however, straightforwardly contradicts the idea of boundedness. This correspondence sets topological equivalence between the projection of every cycle in the (T,S) section and that rectangular whose area and perimeter best fits both its area W and perimeter L - see Figure 1.

The importance of the fit by a rectangular is set by the circumstance that it presents the projection in the (T,S) section of a Carnot cycle that operates reversibly between two effective "heat reservoirs" at "temperatures" T_H and T_L . To remind that a Carnot cycle is a reversible thermodynamical cycle constituted by a sequence of isothermal and isentropic processes; thus it is presented by a rectangular in the (T,S) section.

Further, the perimeter of the original cycle L also contributes to its energy balance because it is straightforwardly related to the energy dissipated by the fluctuations of the interaction system-environment. Indeed, this energy is proportional to the product of the perimeter L and the integrated power spectrum of the fluctuations is proportional to the variance of the time series.

As a result, the approximation of the original cycle by an effective Carnot cycle makes available the evaluation of the efficiency of the tailored engine η through the efficiency of the effective Carnot heat engine η_0 :

Figure 1. Topological equivalence between the projection of every cycle in the (T, S) section and the rectangular section, whose area and perimeter best fits both its area W and perimeter L

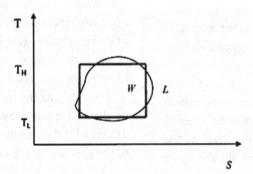

$$\eta = \eta_0 = 1 - \frac{T_L}{T_H} \qquad (9.9)$$

It should be stressed that the ban over degeneration of cycle in $\left(F_i, \Lambda_i \right)$ section into a line in $\left(T, S \right)$ section renders that the efficiency of the effective Carnot engine never to be zero and never to reach 100% !

It is worth noting that the above derivation is free from any dependence either on the operational characteristics of the tailored engine or on the nature of its coupling to the environment. In turn, it fundamentally generalizes the idea of banning the perpetuum mobile because the release from the necessity of physical coupling to two heat reservoirs makes it available for every engine that operates reversibly. To this end, it should be stressed that though the practical efficiency of a tailored engine can be enchanced by appropriate manipulating of the parameters of the cycle, the best efficiency never reaches 100%!

Ban over Information Perpetuum Mobile

It is worth noting that the above ban over perpetuum mobile could be formulated as ban over exerting a non-mechanical work without undergoing changes in homeostasis. Let us suppose that a cycle involves visiting of at least 3 basins-of-attraction. Since the difference in the homeostatic characteristics is $1 - st$ kind discontinuous, there exists an orbit of minimal length such that neither orbit can be made shorter than it by means of a continuous manipulation. Therefore, it is obvious that each semantic unit retains its minimal "engine' so that to render its further manipulation impossible. On the contrary, an orbit that belongs entirely to a single basin-of-attraction could be continuously "shrunk" to a point and thus it does not retain a minimal "engine." Taking into account the fact that such orbit is constituted by random

motion alone, one concludes that it is impossible to construct an information perpetuum mobile, i.e. to transform into semantics a noise alone, i.e. it is impossible to acquire any semantics without involvement of certain changes in the homeostasis.

One of the major advantages of the derived above ban over perpetuum mobile is that it is free from any relation about entropy maximization or extremization of any other state variable. This provides its enormous ubiquity since now the ban over perpetuum mobile is valid not only for simple physical machines but it is spread over the entire majority of complex systems as well. It is worth noting on the highly non-trivial role of the functional organization: both for simple physical systems and their complex counterparts the discontinuous change in functional properties (homeostasis) during the working cycle is that crucial property whlch substantiates the effective Carnot engine which in turn justifies the ban over perpetuum mobile.

Outlining, the enormous ubiquity that the ban over the perpetuum mobile gained by the above derivation justifies its announcement as one of the fundamental laws of Nature and, following the tradition, we proclaim it as the most appropriate and ubiquitous reformulation of the Second Law.

CONCLUSION

The major task of the introduced in the present book concept of boundedness is to create a systematic non-reductionistic approach that is equally available to the behavior of both simple physical systems and the complex ones. The central implement of the non-reductionism is grounded on the understanding that behavior of every system is dictated by its functional organization. To compare, the traditional statistical approach is grounded on the assumption that the behavior of the complex systems is reducible in one way or another to the behavior of the simple systems. This view is very well developed in the traditional statistical thermodynamics where the behavior of the ideal gases and the homogeneous solutions is considered decisive for the behavior of the complex systems as well. This approach led us to the understanding that the Second Law, viewed as a governor of the arrow of time, is one of the basic laws of Nature.

It is to be expected that the radical shift from the reductionistic setting of the traditional approach to the non-reductionistic concept of boundedness will result in radical change of the entire paradigm of the behavior of the natural and man-made systems. A great advantage of the concept of boundedness is that it comprises self-consistently the conditions for holding Maxwell relations. Their importance lies in the fact that they define such important thermodynamical characteristics of every many-body systems such as heat capacity, susceptibility etc. It should be stressed that, though their mathematical form is left intact under the shift of the concept, their understanding is completely new since it explicitly involves the functional organization of the system by means of the role of the equation-of-state.

A crucial advantage of the concept of boundedness is its success in giving a radically novel understanding of the origin of the arrow of time which is completely disintegrated from the notion of entropy. Thus, this eliminates the major obstacle on the road of releasing the association of the entropy with its maximization at equilibrium. In turn, this release makes possible to introduce a completely novel notion of entropy which is equally available for both simple physical systems where the traditional entropy maximization is available and their complex counterparts where the traditional entropy maximization is inherently impossible.

Another fundamental difference between the traditional approach and the concept of boundedness is the explicit involvement of the hierarchical organization in the definition of the notion of internal energy. Indeed, the involvement of the hierarchical organization is made by means of imposing condition for balance between the random and the orbital motion in the state space which stands also as a ban over spontaneous formation of steady fluxes. This condition has a two-fold explanation: (i) it is the implement for substantiating the hierarchical order in the top down direction. The imposing of hierarchical order in the direction *top down* is one of the major distinctive characteristics of the non-reductionism of the concept of boundedness. To compare, the traditional reductionistic approach admits the *down top* direction as the only admissible one; (ii) it puts a ban over "self-propelling" of a non-mechanical "engine" and thus puts a ban over the possibility of creating semantics from the noise alone. This result is strongly supported by our present results about the ban over information perpetuum mobile. Yet, the ban over the information perpetuum mobile has more widespread application since it is universally available: its validity is spread not only over complex systems but over simple physical systems as well.

The ban over the information perpetuum mobile is a counterpart of the most general formulation of the Second Law which asserts that it is impossible to build a non-mechanical "engine" which works steadily in a cyclic regime without exerting any functional changes of its homeostasis during the working cycle.

Chapter 10
Is Semantic Physical?!

ABSTRACT

A critical comparison between the traditional algorithmic approach and the semantic-like one is made. The comparison include topics such as causality, correlations, halting problem, shortest algorithm, intuition, Zipf's law, and absolute information. The purpose of making this comparison is to delineate neatly the fundamental difference between both approaches and to make clear that, although they are different, they still are counterparts which coexist peacefully. One of the major differences between them turns out to be that whilst the semantic-like approach permits autonomous discrimination between "true" and "false" statement by an intelligent complex system, the traditional algorithmic theory does not allow any autonomous discrimination between a "true" and a "false" statement. On the other hand, their common property turns out to be that it is impossible to acquire absolute knowledge: for example, even the famous "super-minded" Maxwell demon can be deceived.

INTRODUCTION

The strategic commitments undertaken within the Kyoto protocol are best highlighted by the initiation of a wide range of activities within each of its objectives. They outline the following assignments that have enormous impact on all aspects of the human activity:

1. Ascertaining a general strategy that breaks the tendency to meet the rapidly growing utilization demands to the human-made products through downsizing the size of elements and their doubling in number because of the extensive growth of the energy costs that accompanies it. This problem gives rise to the question whether there is another, more successful with respect to the energy

DOI: 10.4018/978-1-4666-2202-9.ch010

efficiency, relation between the structure of complex systems and their functioning. The intrigue that lies in the problem is set by the trap between the matter organization governed by the physical laws which, in their traditional understanding, are passive to any form of information and hence are passive to the effectiveness of functioning measured by the created and transmitted information. The problem is that, counter to our intuition and experience, the physical laws do not consider any functional relation between energy and information and thus makes the effectiveness of the functioning passive to the particularities of the applied stimuli and the self-structuring of the system. The critical dilemma is whether the solution of the problem needs fundamental reformulation of the physics or it should be considered as a separate field of engineering science. A decisive argument in favor of the first alternative is that all natural and human-made objects and all living organisms are made of atoms and molecules whose behavior is governed by physical laws. A strong argument in favor of the second alternative comes from the thermodynamics that predicts the work of heat machines and justifies prohibition of the perpetuum mobile but at the expense of considering macroscopic properties described only probabilistically and insensitive to the dynamics of its constituents. The confrontation between these alternatives goes down to the basic principles of thermodynamics because its postulation that the energy is always an extensive variable renders the energy cost of any response fixed to the sum of the energy of the constituents and thus makes it insensitive to the structural organization and its functioning.

2. It is generally expected that without disruptive new technologies, the ever-increasing computing performance (commonly known as Moore's law) and the storage capacity achieved with existing technologies will eventually reach a plateau. At present, the scientific community cannot establish consensus on what type of technology and computer architecture holds most promises to keep up the current pace of progress. However, we comprehend the major move ahead not as the best choice among the variety of contemplated future and emerging technologies (*quantum computers, molecular electronics, nano-electonics, optical computers and quantum-dot cellular automata*) but in establishing grounding principles of a next generation performance strategy that opens the door to realization of a functional circuit capable to autonomous comprehending and creating information.

The major challenge to this idea comes out from the need of establishing one-to-one correspondence between the information and the functional organization of a circuit necessary for making creation of information a controllable process. In turn, the target correspondence requires new understanding of the notion of information which is not reducible to that of Shannon because he considers the information as a quantitative variable which may not distinguish between functionally different states. The obvious requirements to the new understanding of information such as sensitivity to functional morphology, its stability and reproducibility grow into challenges to thermodynamics and kinetics. Thus, the requirement for the reproducibility of information (and the corresponding functional organization) is implemented through reversible transitions between states of different entropy which, however, is in conflict with the Second Law in its traditional understanding. Further, the stability of the functional organization sets certain relation between energy and entropy landscape of the circuit which explicitly relates the information to the energy and thus for the first time defines the energy costs of the information processing. To compare, traditional thermodynamics postulates energy as an extensive variable which renders the energy cost of any response fixed to the sum of the energy of the constituents and hence makes it insensitive to the structural organization and its functioning. Therefore, the conflict between the task of looking for better performance strategy and the grounds of the physical science grows into dilemma whether the physical science must be reformulated so that to make its frame sensitive to information or to consider the problem as a separated field of engineering science. The impact of its solution is enormous since the problem concerns not only human-made circuits but every natural system as well.

Our answer to the above fundamental problems is the developed by us theory of boundedness. Its major assertion is that the physical science must be reformulated so that to make it sensitive to information. The basic steps for meeting this goal come naturally under the mild constraint of boundedness and the rather radically novel understanding of the role of functional organization of a complex system. The most prominent result of the interplay between the constraint of boundedness and the functional organization of any complex system is that its response becomes semantic-like. It must be highlighted that the advantage of the semantic-like response lies in its irreducibility to an algorithmic-like response.

It is worth noting that so far the algorithmic-like response is the dominant understanding of the interactions of any system with its environment. This paradigm consists of assumption that the response is local and after each session of environmental impact a system returns to a stable state called equilibrium. Thus, the major changes in a system follow the "catastrophic scenario", i.e. a major disruptive event happens suddenly, without any signs of warning, and is provoked by large enough environmental impact. Thus, in this setting, the only way of predicting a hazardous event, is the probabilistic one. It is believed that the probabilistic approach to predicting future events

is firmly grounded on the fluctuation-dissipation theorem which in the present context implies that the linear regime provides possibility for studying a system non-disruptively by measuring its linear response. Then, the traditional paradigm assumes that it is possible to gain enough knowledge about a system by means of formal logic, i.e. through imposing an ordered sequence of binary questions (questions which require yes/no answer) and, by means of appropriate measurement, to answer with certainty each of the questions in an algorithm. Nowadays this paradigm is essential implement that is in use in all branches of modern science. The greatest danger comes from the now dominating in all scientific domains (including humanities and social sciences) purpose of achieving an absolute objective knowledge.

This believe is easy to be shattered by the following consideration: let us consider the law about the period of a pendulum $T \approx \sqrt{\dfrac{l}{g}}$. To the question 'whether the period is proportional only to the length of the pendulum' the answer is 'yes', but on the condition that the acceleration is constant. However, the latter is true on the Earth surface only. Thus, as it is well known, every law is derived from certain background conditions; an often overlooked point is that, by taking into consideration the law itself, we cannot judge non-ambiguously about the conditions for its derivation. Consequently, no formulation in terms of yes/no answers is possible for not *a priori* predetermined response. Therefore, the inherent necessity of outside knowledge makes every piece of knowledge situated and relative: even in the case of intelligent binary questions.

Another flaw of the traditional algorithmic approach is related to the probabilistic manner of defining the notion of information: the grounding assumption of Shannon`s definition of information is that both the noise and the information are created by the same stochastic process. Then, the lack of any physical discrimination between the information and the noise gives rise to complementarity between the Shannon information and the entropy viewed as a measure of disorder. In turn, the obtained complementarity makes the information a quantitative characteristic whose value is subject to uncertainty determined by the entropy. The decisive drawback of this duality, however, is that it generates a conflict with the time-translational invariance of the underlining processes as long as the non-zero entropy renders non-zero probability for deviation from any long-term steady behavior regardless to whether the conditions for reproducibility are the same or not. The conflict is best elucidated by the paradox that it generates, namely: the considered complementarity renders the equilibrium, viewed as the state of maximum entropy, to be the state whose behavior would be predicted with the lowest possible accuracy!? Further, the lack of any physical discrimination between the information and the noise highlights the paradox by posing the question whether the Sun will rise tomorrow is also subject to prediction, the accuracy of which depends on the current entropy of the Solar system and its environment!? It is obvious that the core of these paradoxes lies in the parity between the information and the entropy established on the grounds of the oversimplified relation between order and disorder expressed through the dependence of both information and entropy on the probability for realization of a given state. In order to avoid the problem, we assert that the information must be defined so that to account for the functional organization of the "order".

The presented above considerations prompt the major goal of the present Chapter: to delineate the major differences between the algorithmic approach and the semantic-like one as well as the advantages of each of them.

BACKGROUND

The major differences between the algorithmic approach and the semantic-like one could be classified into two groups: (i) the first one is related to their ability for autonomous comprehending and creating of information; (ii) the second one is related to whether their implementation goes via extensivity of the involved processes or they are organized in non-extensive way. The above classification commences from the fact that the semantic-like approach and the algorithmic one are actually counterparts exerted by different physical objects and processes: thus, the only complex systems capable of hierarchical type self-organization are able to produce semantic-like response. The exclusive property of the latter is that each semantic unit is expressed through performance of corresponding non-mechanical engine. In turn, since the performance of any engine is irreducible to the sequence of "letters" by which it is constituted, the corresponding piece of semantics is irreducible to the algorithmic value of the corresponding 'word" viewed as an algorithm. Further, not every sequence of letters constitutes a 'semantic unit' – the physically allowed "words" are only those consistent with the boundedness. Moreover, an exclusive property of hierarchically self-organized complex systems is that their semantic-like response is spontaneous, i.e. it appears as a result of exerting natural processes. On the other hand, the hardware of every algorithmic approach needs outside involvement since after each logical step, it must be re-set to neutral position so that all states to be available for the next logical step. Thus, since the re-set is accomplishable only through hand-craft influence, the algorithmic approach is not able to create autonomously information. Further, there is not a natural mechanism, similar to performance of engine, which would select a meaningful sequence of "letters" from a random one. The positive aspect of the algorithmic approach is that in general its non-autonomy allows that any external mind is capable to discriminate among several "hypothesis", each grounded on formal logic. Although the option of such discrimination is open, its result should not be taken as an absolute truth since our minds are also "built" of atoms and molecules and thus we are open to misjudging.

On the contrary, the semantic-like response "tells" us the best way of adapting the functional and/or structural organization of a complex system so that its behavior to be the most stable in the current environment. Thus, the knowledge we acquire from a semantic-like response is always situated. Nonetheless, though situated, the knowledge is still objective since identical functional organization will result is the same piece of knowledge. The exclusive property of that knowledge is its autonomous comprehending: agents of identical functional organization when being subjected to identical impact comprehend the same knowledge independently from one another.

The independent comprehending of identical knowledge poses the fundamental question how the agents recognize the signal from the noise. The problem is particularly acute for the systems of complex functional organization where the signal and the accompanied noise that comes from the hierarchical feedbacks are distorted non-linearly by the interactions among different nodes in every hierarchical level. The problem may have another interpretation: is it ever possible to "code" a message so that to transmit it safely through a noisy channel without changing the cipher? Thus we face the long-standing problem about the secure exchange of information which becomes more and more serious in view of rapidly growing communications which use noisy public channels such as Internet, telephones etc. The problem has two major aspects that are particularly acute because of their immense social impact: on the one hand, it is necessary to diminish as much as possible the distortions of the information provoked by the contingencies of the traffic noise and on the other hand to protect it from intentional attacks on its privacy. So far, these problems have been treated predominantly separately. Although a remarkable success has been achieved and a number of strategies have been developed, the solution of these problems still lacks general approach, insensitive to the particularities of coding and ciphering (or their combination). The weak point

of all developed so far strategies is the vulnerability of the encrypting protocol to the noise since the latter can essentially distort the encapsulated information. We assert that by means of the same protocol the information can be made both robust to the noise and secure to intentional attacks on its privacy. The solution lies in the properties of the proposed by us new definition of information – discrete band in the power spectrum of a bounded time series. As we have already proved in Chapter 2 the general condition for providing long-term stability, the boundedness, ensures the insensitivity of the information to the noise statistics, i.e. whatever the statistics of the noise is, the information encapsulated in the discrete band can be obtained with the same accuracy over and over again. This result proves the robustness of the information to the noise in the general frame of boundedness as a single constraint imposed on the structure of the bounded time series. The next step is to demonstrate that the same circumstance helps for considerable enhancing of the robustness to intentional attacks on the privacy of the information. Indeed, by appropriate embedding of the information, viewed as a "homeostatic"-type process, into bounded noise with arbitrary statistics we can produce a bounded irregular time series that can be safely transmitted through public noisy channel. The major advantage of this approach is that the sender can change the statistics of the noise unilaterally and the receiver(s) still can decrypt the information encapsulated in the discrete band. The unilateral change of embedding noise statistics is a new powerful tool for fast reaction to any contingency and to any attempt to break the code and/or to compromise the key. The strategic benefit of the obtained unilaterality is that its use considerably reduces the need of permanent complicated multilateral synchronization of encryption and decryption keys persistent even with the use of asymmetric ones. In addition, it is worth noting that the "security through embedding into noise" is an entirely novel approach whose major property is that it is generic for all types of cryptographic protocols regardless to the particularities of the code, cipher and key used. It should be stressed that relevance of the use of "homeostatic"-type processes for the "alphabet" is grounded on their exclusive property to be insensitive to the details of the noise statistics. Further, as it would be demonstrated later in this chapter in section *Boundedness in Reality*, their embedding into noise covers their characteristics so well that it is impossible to judge about the change of their characteristics on the grounds of the time series only.

It is important to stress that different hardware produces different response: we already pointed out that only complex systems that are able to hierarchical self-organization can produce semantic-like response. On the other hand, the most elaborate hardware that produces algorithmic approach is the Boolean networks. To remind, Random Boolean Networks (RBN) are known as NK networks or Kauffman networks. An RBN is a system of N binary-state nodes (representing genes) with K inputs to each node representing regulatory mechanisms. The two states (on/off) represent respectively, the status of a gene being active or inactive. The variable K is typically held constant, but it can also be varied across all genes, making it a set of integers instead of a single integer. In the simplest case each gene is assigned, at random, K regulatory inputs from among the N genes, and one of the possible Boolean functions of K inputs. This gives a random sample of the possible ensembles of the NK networks. The state of a network at any point in time is given by the current states of all N genes. Thus the state space of any such network is 2^N. The major property of every RBN is that, on fine-tuning of the input, it reaches an attractor where it stays for arbitrary long time. The remarkable property of that state is its robustness to local failures. In turn the latter makes it very attractive for developing a promising hardware which is insensitive to local failures. This is an example how a network when its elements (nodes) are being subjected to formal logic reproduces this property at the next

level. Indeed, each node could reside in one of two states (active or inactive) and switches between them under the command "IF": "IF" the neighborhood impact reaches certain threshold, the node switches its state. The same is with the attractors: "IF" the local concentration of the nodes in certain state reaches certain value, the system switches to another attractor. Thus, the Boolean networks indeed reproduce formal logic in a way that makes them much more diverse in their abilities and much more reliable in their operations, properties which render them an elaborate hardware for Turing machines.

The importance of the Boolean networks for us lies in the following: the invariance of the formal logic as a type of response which appears under self-organization is remarkable! Yet, it should be stressed on the fact that the self-organization of the Boolean networks is not a hierarchical one!

COMPARISON BETWEEN ALGORITHMIC AND SEMANTIC-LIKE APPROACH

The goal of the present section is to present the differences between the algorithmic and the semantic-like approach from the viewpoint of the traditional algorithmic theory. Thus, we will consider the difference between both approaches with regard to matters such as causality, correlations, halting problem, shortest algorithm, intuition, Zipf's law, absolute information. The purpose for making this comparison is to delineate neatly the fundamental difference between both approaches and to make clear that, although they are different, they still are counterparts which coexist peacefully.

Causality vs. Correlations

Let us start with the major delineation between the semantic-like approach and the algorithmic one: the boundedness, viewed as general concept of the self-organized semantics, sets limited choice of admissible steps for each future event on the contrary to the grounding assumption of the algorithmic theory which consists of sustaining the same set of admissible choices for every logical step. Thus, the boundedness gives rise to inevitable specific correlations among distant events. The question is whether these correlations are causal? The problem becomes very serious in view of the fact that the sustaining of the same set of admissible choices for the algorithmic approach implies that the distinguishing of any correlations is subject to the expertise of an external conscious mind only. Thus, we face the long-standing debate whether and when correlations are to be considered result of causal relations. Though the debate on this matter is still open, most of the authors are inclined to share the view that the causality must be associated with time-asymmetric binary relations among events. This view corresponds to the suggested by our daily experience observation that if ' A causes B ', A appears before B in the time course but not vice versa. This view has an enormous impact on the methodology of science since it is utilized as grounding assumption for the algorithmic theory which asserts that: an event can be studied by means of answering a specific for the case ordered sequence of yes/no questions. Thus, the time ordering of cause-effect phenomenon appears as grounds of formal logic.

However, the theory of boundedness offers a fundamentally new alternative to this concept: it launches a new type of causality, called by us semantic causality, which is to be associated with the functional irreversibility of the performance of the 'engine' associated with every semantic unit. It should be stressed that each and every such 'engine' involves algorithmically non-reachable one from another states. The crucial for the semantic causality fact is that any engine performs differently if being run in the opposite direction. For example, the traditional Carnot engine operates as a pump in one direction, while in the opposite it performs as a refrigerator. By means of employing the irreversibility of engine performance with

causality, we rather relate causality to a process than with binary time-asymmetric correlations as the traditional view supposes. The association of causality with the functional irreversibility of a Carnot 'engine' provides a way of implementing logical irreversibility as a generic property of any semantic unit.

Alongside, the association of causality with an irreversible process specifies the way how the environmental changes are involved in the realization of any specific correlation among them. Note, a correlation itself is insensitive to the process of its realization and thus it is insensitive to the functional organization of a system that is subject to any given environmental impact. It should be stressed on the fact that the association of causality with the performance of an engine meets self-consistently the requirement about its reproducibility on reoccurrence of the event. Indeed, this way we associate causality with the steady specific homeostatic functioning and structure than with the current specific response which has universal (non-specific) properties. In result, we have more: indeed by means of association of causality with the performance of an engine, we achieve holding causality in an ever-changing environment. It is worth reminding that the traditional notion of causality implies re-occurrence of a causal event only when the same environmental conditions reoccur.

Hierarchy of Semantic Causality vs. Directed Acyclic Graph of Formal Logic

An exclusive property of semantics is its hierarchical organization: thus, letters are organized into words, words into sentences, sentences into paragraphs etc. On the other hand, the formal logic is organized into acyclic directed graph and so it does not retain any hierarchy. The organization into acyclic directed graphs implements the major property of the formal logic to be constituted by an ordered sequence of binary questions.

This is the way how the formal logic meets the fundamental observation that if 'A causes B', A appears before B in the time course but not vice versa. The weak points of the presentation of causality via directed acyclic graphs are: (i) it is not automatically guaranteed that all paths of that graph comprise processes that are in balance with the environment. Put in other words, it is possible that some paths are constituted by processes that are in balance with the environment while other do not. In result, it is not certain *a priori* whether a given path (given piece of formal logic) is physically available or not. To remind, when a given path is not in permanent balance with its environment, the corresponding system would collapse.

In contrast with the formal logic, the semantic-like response is organized into cycles whose performance is hierarchically subordinated. The hierarchical subordination is implemented by means of stationarity of 'action' which makes a given "engine" to perform in a specific direction. It is worth noting that the coordination of the 'engines' performance automatically provides a stable functioning of any system and thus the latter is in permanent balance with its environment. It is worth noting that the hierarchical subordination into cycles that operates as 'engines' provides automatic discrimination between a random sequence of semantic units and a meaningful one. Indeed, a random sequence of semantic units is not a subject of hierarchical subordination while a meaningful one does. A prominent example of this statement is the famous Zipf's law. To remind, it asserts that given some corpus of natural languages, the frequency of any word is inversely proportional to its rank in the frequency table. Thus the most frequent word will occur approximately twice often as the second most frequent word, three times as often as the third most frequent word, etc. Put in other words, the Zipf law ignores any semantic meaning and thus seems to sweep out the difference between mind activity and random sequences of letters. On the other hand, it is a common knowledge that semantics is permutation sensitive and puts a long-range order among its units. Then, how is it possible to occur that these long-range correlations are ignorant? The answer is not easy and comes only after a closer look at the structure of a semantic sequence. We start with reminding that under a concept of boundedness, any semantic sequence is a sequence of orbits in the corresponding state space. The permutation sensitivity is implemented by associating the meaning of each orbit with the functional irreversibility of a specific engine. Thereby, by means of imposing certain hierarchy in the running of the orbits, i.e. running each in a specific direction, specific long-range correlations are substantiated as set by semantics. However, when considering the frequency of occurrence of a given word in a text, we ignore these higher level correlations and reduce the "system" to its simplest counterpart, i.e. to the first hierarchical level; and as we already have established in Chapter 5, its state space is a dense transitive set of orbits subject to power law distribution. Thus, the hierarchical order of 'engines' performance gives rise to the hierarchy of semantic causality. Yet, perhaps, the most amazing property introduced by the hierarchy of semantics is that it opens the door to the uniqueness of any piece of semantics, i.e. the more complex the hierarchy, the corresponding piece of semantics has more unique properties.

True, False, and Improvable Statements: Counterfactuals

A fundamental enigma of the algorithmic theory and the scientific methodology is the problem how to discriminate between a 'true' and a 'false' statement; how to prove a given statement with certainty; are the counterfactuals true or false statements. The importance of this matter enormously increases in the present days when the human activity becomes more and more intense and when the scientific community conveys more and more relations between different events and systems.

The traditional scientific methodology is grounded on the formal logic, i.e. on the idea that every piece of knowledge is decomposable to a sequence of yes/no answer questions so that the answer to each question is to be defined with certainty by appropriate non-destructive measurement. What is tacitly presupposed in this assertion is that the decomposability of the behavior of a system into a sequence of events implies that the causality is local and such that follow the local response. However, this immediately triggers a vicious circle: if the causality is local, then it is expressed by the correlations between the successive local responses. This, however, is in fundamental conflict both with the idea of equilibrium as global attractor and with the idea of homeostasis as a self-organized pattern robust to small environmental variations.

In order to overcome the difficulty, we put forward the assumption about semantic causality: i.e. causality which is organized as performance of an engine which involves algorithmically non-reachable from one another state. To remind, the concept of boundedness makes different patterns of homeostasis to comprise a specific irrational frequency which in turn makes the corresponding patterns algorithmically non-reachable. At the same time different patterns are physically reachable, i.e. reaching admissible state involves finite amount of energy, matter, entropy etc. Taking into account the ban over information perpetuum mobile, i.e. the ban of acquiring information from the noise only, one concludes that in the setting of semantic causality there is a ban over matching between the causality and the correlations. Put it in other words, semantic causality requires algorithmic unreachability of its subjects. The withdrawal from any relation between the correlations and the causality provides the ability to discriminate between the 'true' and 'false' statement and to give a clear idea what counterfactuals are. Indeed, the fundamental difference between the semantic causality and the formal logic is that the semantic causality selects only those trajectories

in the state space which sustain permanent balance with the environment. Then, a 'true' statement corresponds to that unique trajectory, which has been physically realized in a given run; a 'false' statement is the one that follows an imaginary trajectory that comprises a non-admissible state; a counterfactual corresponds to an admissible trajectory which has not been realized in a given run. And last but not least, an improbable statement corresponds to an imaginary impasse: the statement does not correspond to any trajectory. Thus, the automatic selection of trajectories to admissible and non-admissible renders the fundamental difference of the present approach from the traditional algorithmic theory and the formal logic which are grounded on the assumption that all trajectories are available.

As a general example, the manipulation theory of scientific explanation (Woodward, 2003) corresponds to the operational protocol of behavior of complex systems. Thus, an intelligent-like behavior of a complex system comprises a high degree of autonomy in creating local hierarchies of 'true', 'false' and 'counterfactual' information units. The higher the autonomy of the systems the more "unpredictable' their choices may be that drive systems both to stability or transformation.

Outlining, it is important to stress that the above classification of true, false and counterfactuals is rendered possible exclusively by means of automatic selection of admissible trajectories, i.e. by the property of a system to self-organize its response so that to remain stable in an ever-changing environment.

Halting Problem

Halting problem is one of the greatest mysteries in mathematics since it is related to its very core: it is related to the notion of irrational numbers and their comparison with the real numbers (the natural and the rational ones). The problem is very serious for computing and is provoked by utilization of the irrational numbers through their presentation in the same metric way as the real numbers. The fundamental flaw of this presentation is that neither irrational number can be reached by a real one in algorithmic way, i.e. by a finite sequence of mathematical operations. Taking into account that the present day computers implement linear processes, it is obvious that a computer "halts" whenever it should compute an irrational number. Yet, the problem is that one cannot *a priori* know whether a given algorithm would halt or not. And this famous and very important result comes out from the property of the recursive mathematical operations to result either in real and/or in irrational numbers when are proceeded numerically. Thus, for example, the operation 'square root' applied to the natural number 2 gives an irrational number but when applied to the natural number 4 gives the natural number 2. Thus, it is obvious that even though a given relation can be expressed in a closed recursive form, its numerical computation could produce at least one irrational number. The fundamental result is that it is impossible *a priori* to judge whether a given algorithm would halt or not.

The halting problem is rather a fundamental problem of mathematics than a technical problem how to proceed with the truncation when face computing of an irrational number. The importance of the problem lies in its relation with the widespread belief that the laws of Nature could be expressed in a recursive way which would serve as grounds for asserting that the Universe is a Turing machine. Then we face the following difficulties: the expression in a recursive way of the Laws of Nature implies that they are algorithmically reachable one from another. Further, the tacitly presupposed assumption that the irrational numbers share the same metrics as the real ones makes the physical reachability of any two algorithmic different states subject to "dice throwing", i.e. whether the distance between given states is real or irrational. However, if the distance is expressed through an irrational number, the algorithmic reaching the corresponding

"law" would take infinite number of steps. Put in other words, some "laws" would be algorithmically "reachable" whereas some other would not. The greatest paradox would be that it would be *a priori* impossible to judge whether a given "law" is reachable or not. Moreover, applying different algorithms, we can obtain that a given law is reachable by means of certain algorithms while by means of the others the "algorithm" "halts".

In order to overcome the problematic considered above, we presented a radically novel approach to the notion of irrational numbers. In chapter 4 we suggest their introduction in a metric-free from. To remind, we assign the property of "irrationality" to specific lines in the discrete part of the power spectrum of bounded time series. These lines come out from the property of the power spectrum of a self-organized system to be additively decomposable to a discrete band at appropriate control parameters, and a continuous band of shape $1/f^{\alpha(f)}$, which comes from the corresponding inter-level feedback. The fact that both bands in the power spectrum come from the same equations of self-organization is justified by the presence of a non-recursive line coming from their interplay. The interplay commences from the non-linear and non-homogeneous way of participation of the homeostatic and inter-level feedback components in the corresponding equation-of-state. And since the overall solution of the equation-of-state is bounded irregular function, it could be presented as wild and permanent bounded "twisting" and "winding up" around a "skeleton" built on the discrete band of the power spectrum. Thus, the motion resembles the motion on a torus with irrational frequencies. In the present case the motion never stops and thus extra-line associated with irregular motion around the skeleton appears always as non-recursive even-though the thresholds over the corresponding fluctuations are free to be in any relation, recursive one included, with the other scales in system. Thus, the non-recursive component substantiates ban over resonances between "noise" and "homeostasis" viewed as a condition of maintenance of permanent boundedness.

The greatest advantage of the above metric-free way of defining irrational numbers is that it makes different homeostatic states algorithmically unreachable from one another yet, at the same time, they are physically reachable. In turn, this eliminates the halting problem since all states become physical reachable by means of involving finite amount of matter/energy and time. Thus, semantic-like computing is free from the halting problem.

Another important advantage of the considered novel viewpoint on the irrational numbers is that different laws of Nature viewed as different "homeostatic" patterns stand as algorithmically unreachable one from another. It comes to provide a novel view on one of the most fundamental problems in the information theory, that is to what extend the information and its processing is related to the organization and functioning of the natural processes. The enormous development of the information theory culminates in the recently widely discussed matter whether the laws of the Universe are computable. The extreme importance of that matter is provoked by the universality of Turing machines which comes to say that every law, viewed as a quantified relation between certain variables, expressed in a closed form, is computable. Then, an immediate outcome of the traditional algorithmic theory is that the variety of laws turns compressible (reducible) to its shortest form so that there is no shorter algorithm executable on any Turing machine. In turn, the existence of universal algorithm (shortest program) whose major property is its insensitivity to the hardware (Turing machine) implies that our Universe is subject to universal law so that each of its constituents obeys it irrespectively to its specific properties. Thus the algorithmic compressibility smears out any difference between one specific property and another.

Therefore, the algorithmic unreachability of each law from any other provides the huge diversity of laws in the Nature and their uniqueness.

SHORTEST ALGORITHM VS. "INTUITION"

One of the major goals of the algorithmic information theory is establishing of general criteria for discriminating a random sequence of symbols from a sequence that has certain meaning in a given language. So far the scientific community shares the view that such general criterion is given by the so called Kolmogorov complexity: Kolmogorov defined the complexity of a string x as the length of its shortest description p on a universal Turing machine U. A string is simple if it can be described by a short program, like "the string of one million ones", and is complex if there is no such short description, like for a random string whose shortest description is specifying it bit-by-bit. Kolmogorov complexity (K) is a key concept in (algorithmic) information theory. An important property of K is that it is nearly independent of the choice of U. Furthermore it leads to shorter codes than any other effective code. K shares many properties with Shannon's entropy (information measure) S, but K is superior to S in many respects. To be brief, K is an excellent universal complexity measure, suitable e.g. for quantifying Occam's razor. The major drawback of K as complexity measure is its incomputability.

An overlooked so far property of the K complexity is its explicit relation to the metric properties of the corresponding sequence. To make it clear let us consider the role of the K-entropy as time series invariant: it puts universal unavoidable limit over predictability of the future behavior of each *BIS*. Thus the behavior of a BIS drives a vicious circle: (i) on the one hand, its detailed future behavior is unpredictable: (ii) on the other hand, since the behavior of a BIS is additively decomposable to a steady "homeostatic" part and a universal "noise" one can conclude with certainty whether this behavior will be sustained in the future. The solution of the problematic lies in the different account for the metric properties of the terms in the sequence which presented the *BIS* : unpredictability of the future behavior of *BIS* considered as a metric property is grounded on establishing and operating with specific distances among the members of the time series. Here we encounter the same situation as with the halting problem - one never knows *a priori* whether a program computes real or natural number and this uncertainty is grounded on the metrics-based definition of enumerating real and irrational numbers. Thus, finiteness of K-entropy once again sets the halting problem as a problem of metrics-based distinguishing between the natural and real numbers. On the other hand, the additive decomposability of the power spectrum to a discrete and a continuous band is an exclusive property of boundedness which allows a metric-free form of defining irrational numbers and thus serves as grounds for irreducibility of the description of a BIS 'bit-by-bit' as prescribed by Kolmogorov.

The description of a *BIS* through its power spectrum is the expression of the property of self-organization substantiated by extended both in the space and in the time patterns whose description thus becomes non-local and metric-free. On the contrary, the Kolmogorov complexity is defined as a property based on the local metrics. This difference has far going consequences: (i) the Kolmogorov complexity cannot serve as a criterion for applying Occam's razor since there may be shorter description (metric-free one); (ii) since the metric-free description is always associated with certain hierarchical order, it could be named as "intuition". Here we consider the notion of intuition as finding a shorter description at higher hierarchical level than the shortest one at any given hierarchical level. Thus, the property of inter-level feedbacks that provides the

presence of irrational lines in the corresponding power spectra has a very far going consequence: it opens the door to the notion of intuition viewed as establishing of a shorter description at higher hierarchical levels.

No Absolute Knowledge

One of the major problems of the scientific methodology is the issue about the certainty of verification of any theory. Nowadays it is generally accepted that the only path to verification of a given theory goes through an appropriately set experiment. The issue gains greater importance in view of rapidly growing complexification of the experimental equipment and the involvement of more and more assumptions for reading the obtained results. A systematically overlooked point is whether the results could be inherently uncertain, even if we know and/or guess the reasons for their uncertainty. If so, it is obvious that the experiment could not be considered as an undisputable argument for verification or rejection of any theory; it should be considered only as one of the arguments.

In order to make ourselves clear, we will consider how the famous explanatory problem known as Maxwell demon where the 'mind of the demon' could be deceived by his/her own reasoning that involves the non-unitary interactions introduced by us in Chapter 3. 140 years ago the famous Scottish physicist James Maxwell supposed that some kind of entity (a "demon") could be invented to act as a gatekeeper between two isolated chambers of gas, letting only fast molecules into one chamber and only slow molecules into the other. In doing so, he proposed, the difference in temperature between the two chambers would progressively increase, thus violating the second law of thermodynamics. The flaw in Maxwell's thinking, however, was that the demon itself would expend energy when controlling the gate thus saving the Second Law from violation.

Recently, Maxwell demon attains central role in the long-standing debate about the relation between the Second Law and the physical implementation of the notion of information. The following list of excellent papers highlights the subject-matter best: (Bennet, 2003), (Parker&Walker, 2004), (Zander, et al, 2009), (Norton, 2005), (Shenker, 2000) and (Plenio&Vitelli, 2001)

Our task now is to consider the Maxwell demon from different perspective: during his invention 140 years ago the intellectual power of the demon has not been disputable. However, now we pose the question whether the demon could unmistakably separate the molecules according to their velocities when taking into account the non-unitary interactions. To remind, in chapter 3, we introduce the non-unitary interactions as a natural implement for sustaining the boundedness of the macroscopic variables such as concentrations. The non-unitary interactions are defined as follows: along with the unitary processes there always are non-unitary processes which come out as a result of the inevitable involvement of the random motion of the molecules. Indeed, any collision could be interrupted by another molecule that enters the interaction as a result of its random motion. The non-unitarity is due to the time-reversed *asymmetry* between the ingoing and outgoing trajectories of the colliding molecules which commences from the dependence of the outgoing trajectory on the status of interaction at the moment of arrival of the new molecule. A crucial for our considerations property of the non-unitary interactions is that they are metric-free, i.e. the elements of the non-symmetric matrix that represent the Hamiltonian are random but bounded and of the same order. The property of being random is to be associated with the interruption of an unitary collision by another molecule at a random moment; and since the original distance to the interrupting molecules is of no importance for any given collision, a non-unitary interaction is not specifiable by metric properties such as the distance between molecules - it is rather specifiable by the correla-

tions among the interacting molecules so that the intensity of these correlations are permanently kept bounded within metric-free margins. Therefore, each act of measuring velocity of a molecule, the measurement could be interrupted by entering another molecule in the colliding process which implements the measurement. The crucial facts are two: (i) the first one is that the probability for interruption of the measuring process is proportional to the concentration of molecules; (ii) the velocity of a target molecule becomes undefined after undergoing a non-unitary interaction. Indeed, the energy and momentum conservation laws are not enough to provide non-ambiguously the velocities of the molecules after the interaction. For example, the energy and momentum conservation laws of a 3-body interaction provide 4 equations for defining 9 components of the out-going velocities. In result, Maxwell demon cannot decide with certainty whether the target molecule is hotter or colder. It is worth noting that the knowledge about the existence of non-unitary interactions does not help the demon in his task. Thus, we conclude, that there is not absolute knowledge and that sometimes even the knowledge itself cannot help judging about an experiment with certainty.

BOUNDEDNESS IN AN EXPERIMENT

So far our attention has been focused on the development of self-consistent theory that accounts for the long-term stability of the systems that exert fluctuations and their exclusive property to exert its response in semantic-like manner. Despite the plausibility of our approach, it would be a mere speculation if not confirmed by the reality. That is why the aim of the present section is to outline how the developed theory can be validated.

The approach developed so far is built on the grounds of advanced by us concept of boundedness which has 3 major aspects. The first one asserts that every system has its thresholds of stability that if exceeded makes a system to undergo breakdown.

This assertion imposes limitation on the amplitude of fluctuations: in order to sustain long-term stability, the fluctuations are exerted so that never to exceed the thresholds of stability. However, at this point our intuition runs into problem: what happens when a fluctuation reaches them - do "U-turns" involve specific physical processes? Our answer is no, no physical process is needed for executing the "U-turns" when an intriguing interplay of the short-ranged dynamics and the chaoticity of the state space takes place. It is explicitly expressed by the relation (5.13) that sets the largest amplitude of fluctuations so that the corresponding "U-turns" occurs without involving any additional physical process. Though the relation (5.13) has been rigorously established, its key role in the entire approach poses the question whether this rather abstract result can be verified in reality. To the most surprise the answer is positive: it indeed can be verified. Let us recall that the assumption about the automatic undergoing of "U-turns" renders scale invariance of the time scales correlations, i.e. all time scales contribute uniformly to the evolution. We will elucidate this issue in the next sub-section where the effects of the interplay between the scaling invariance and the boundedness are considered. Moreover, we will present a broad spectrum of experiments that support them.

The second specification of the basic concept is the dynamical boundedness. It originates from the fundamental assumption that every process in the Nature develops so that the rate of exchanged with the environment energy and/or matter is always bounded. The formal explication of that assumption implies boundedness of the rate of fluctuation development. Evidently, it sets relation between the size of a fluctuation and its duration. Furthermore, as proven in the section *Boundedness of the Rates. Embedding dimension* in chapter 1 the parity between the parameterization through the phase space angle and the duration of the fluctuation renders every fluctuation to have its own embedding dimension.

Hence, the embedding dimension of a time series varies with its length because its current value is set on the size of the largest in the time series fluctuation. On exerting the first "U-turn" the embedding dimension reaches its largest value and stops varying on further increase of the length of time series. Further in the present section we will consider concrete experiment that manifests variable embedding dimension.

The third specification of our fundamental assumption is that of hierarchical organization. The hierarchical organization is considered as an implement for strengthening of response in an ever-changing environment. Alongside, the hierarchical-like order appears as generic implement for sustaining homeostasis by means of substantiating intra- and inter-level interactions through specific feedbacks which operate in semantic-like manner. Therefore the diversity of properties is reached by diversification of the hierarchical structuring. The wisdom of exerting inter-level feedbacks in a form of bounded noise lies in the proved in Chapter 1 insensitivity of the shape of the noise band to the statistics of the noise. Further, this fact, along with the additivity of the noise and the discrete band in the power spectra, provides robustness of homeostasis to the details in environmental variations. The hallmark of the boundedness of the noise that implements an inter-level feedback is manifested through the presence of an irrational frequency in the power spectrum of non-zero discrete part.

An important outcome of the joint action of all 3 aspects of boundedness is that the macroscopic evolution is described by bounded stochastic ordinary or partial differential equations of the type (4.1) and (7.13). The major exclusive property of the solution of these equations is that its power spectrum comprises additively discrete band which characterizes the corresponding homeostatic pattern and a continuous band of shape $1/f^{\alpha(f)}$ that originates from the inter-level feedback. It should be stressed that while the discrete band emerges only for certain choices of the control parameters, the continuous band persists at every admissible choice of the control parameters. On the contrary, neither system of ordinary or partial differential/difference equations gives rise to coexisting of a discrete and a continuous band in the power spectra of their solutions at any control parameter choice. Hence, the coexisting of a discrete band and continuous band of shape $1/f^{\alpha(f)}$ is a crucial test for the validity of the entire theory.

Before presenting the experiment let us recall how the boundedness of the fluctuation size is interrelated with the scaling invariance. To remind, the major characteristics of the correlations among time scales is the autocorrelation function:

$$G(\eta) = \frac{1}{T} \int_0^T X(t+\eta) X(t)\, dt \qquad (10.1)$$

The autocorrelation function $G(\eta)$ is a measure for the average correlation among any two points in a time series separated by time interval η. Yet, more popular is the power spectrum, i.e. the Fourier transform of $G(\eta)$:

$$S(f) = \lim_{T \to \infty} \frac{1}{T} \int_0^T G(\eta) \exp(i\eta f)\, d\eta \qquad (10.2)$$

The power spectrum is due its popularity to its easy and non-ambiguous reading. Indeed, any long-range correlation between two time scales appears as a single line whose amplitude is proportional to their correlation. Therefore, it is to be expected that the scale invariance does not signal out any perceptible line in the power spectrum. In Chapter 1 we found out that the power spectrum of a time series of length T that is subject both to boundedness and scale invariance is continuous band of shape $1/f^{\alpha(f)}$ where $\alpha(f) = 1 + \kappa\left(f - \frac{1}{T}\right)$, $f \in \left[\frac{1}{T}, \infty\right)$. Note that

the non-zero correlations among time scales are introduced by the boundedness and are not a result of any physical process! Besides, if the time series were unbounded, the scale invariance would result in zero correlations among the time scales, i.e. all components of the power spectrum would be zero! Therefore, the interplay of the boundedness and the scale invariance gives rise to continuous band of the shape $1/f^{\alpha(f)}$. So, the persistence and smoothness of that band serve as a criterion for the validity of the assumption about the joint action of the boundedness and the scale invariance. Now we are ready to interpret the experiments. Let me start with a particular one. The purpose is to support as much as possible aspects of our theory by examination of the same experiment. It is expected that since all aspects of the basic concept are inseparably entangled, they are included in one way or another in the same experiment.

The Experiment

We have studied the reaction of oxidation of $HCOOH$ (formic acid) over two modifications of supported Pd catalyst at steady flow of the reactants (Koleva et al, 2000). The control parameters were the partial pressure of the reactants and the temperature of the feedstock at the reactor inlet. The temporal behavior was studied in the temperature interval $110 - 130\ ^\circ C$; the O_2 feed concentration was varied from $0.5 - 12\%$ while that of $HCOOH$ - from 0.5% to 8%.

The difference ΔT between the catalyst bed temperature and the feedstock at the reactor inlet was measured, digitized and continuously recorded. The sampling rate was 2 points per second. 80 time series of that difference have been recorded scanning the values of the feed concentrations and temperature of the feedstock at two charges of the catalyst.

At all 80 time series the difference ΔT exhibits irregular variations the amplitude of which does not exceed 10% of the average "shift" of the catalyst bed temperature from the feedstock one. However, there are occasional large variations amplitude of which about 50% of the average "shift".

Our study shows that all 80 power spectra comprise continuous band of the shape $1/f^{\alpha(f)}$, where $\alpha(f) \to 1$ at $f \to 1/T$ (T is the length of the time series) and $\alpha(f)$ linearly increases on f increasing. The shape is robust to the catalyst charge, feedstock temperature and the feed concentrations. Next we presented as an example two of those time series along with their power spectra (Figure 1 and Figure 2). The time series are presented in digital units and the power spectra in lg-lg scale in relative units: the value of the successive components is divided to the first one. This is made in order to make obvious that the power spectrum is power function of shape $1/f^{\alpha(f)}$ with $\alpha(f) \to 1$ at $f \to 1/T$ (T is the length of the time series) and $\alpha(f)$ linearly increasing on the increase of the frequency.

The power spectra in Figure 1b and Figure 2b show that they indeed fit the shape $1/f^{\alpha(f)}$ with $\alpha(f) \to 1$ for the infrared edge (the first component) and $\alpha(f)$ linearly increase on the increase of the frequency. It should be stressed that the use of relative units for presentation of the power spectrum makes revealing of the exponent $\alpha(f)$ non-ambiguous because it makes the shape $1/f^{\alpha(f)}$ insensitive to the choice of the units. Be aware that the shape $1/f^{\alpha(f)}$ is not unit-invariant! The latter implies that on any change of the units $f' = cf$, the shape $1/f^{\alpha(f)}$ is multiplied by factor $c^{\alpha(f)}$ which, however, is different for different frequencies! Hence, $\alpha(f')$ deviates from $\alpha(f)$! The only way to make $\alpha(f)$ unit-invariant is to present the power spectrum in relative units.

Though the above experiment is very convincing, it poses the question whether it is only a particular case. Remember that we have introduced the boundedness as a concept advantageous for a broad spectrum of systems. Correspondingly, credible support must be brought about by a variety of systems of different nature. Such support does exist and brings about one of the most ubiquitous phenomena in the world: $1/f$ noise. Its major characteristic is that the infrared edge of the power spectra uniformly fits the shape $1/f$. The fit does not depend on: (I) the incremental statistics, i.e. the details of the irregularity succession in the time series; (ii) the length of the time series; (iii) the nature of the system - the phenomenon is observed in large variety of systems: quasar pulsations, meteorology, financial time series, music and speech etc. Though it has been thoroughly studied for more than a century, it still remains enigma. In section *Power Spectrum* in Chapter 1 we have already discussed the relation between the shape $1/f$ and $1/f^{\alpha(f)}$. The purpose to come again to that topic is that since our conjecture it is able to clarify reliably the above listed properties, the proximity of the shapes $1/f$ and $1/f^{\alpha(f)}$ is a strong argument in its favor. Another argument for our conjecture is the irrationality of the farthest left component: even naked eye could judge that it does not belong to the band constituted by all other members of the discrete band.

The advantage of the boundedness conjecture is best revealed in the reconciliation of the greatest mystery of the $1/f$ noise: on the one hand, the shape $1/f$ should be associated with instability and breakdown because it renders the variance of the time series, calculated on its grounds, infinite. In turn, the infinity of the variance implies that fluctuations large enough to carry the system beyond the thresholds of stability become most probable. Therefore, any system would rapidly

Figure 1. A) First time series; B) The power spectrum of the time series presented in 1a

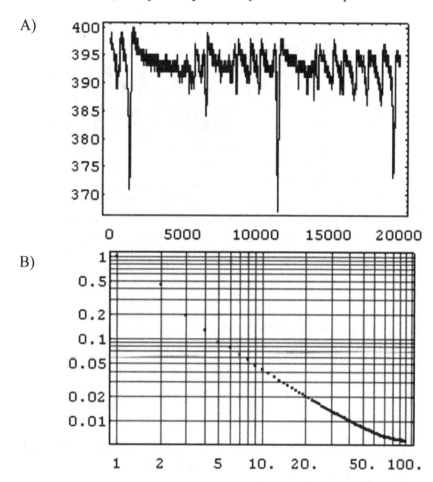

blow up or get extinct. However, it does not happen. Moreover, it is well established that the $1/f$ behavior is spanned over several dozens of orders in the time course. Hence, it should be rather associated with long-term stability. In turn, the latter makes the concept of boundedness advantageous for the $1/f$ noise. The boundedness always sets finite variance of the fluctuations that opposes the infinite variance set on the shape $1/f$ of the power spectra. The considerations in section *Power Spectrum* and section *Variance. Setting of k* in Chapter 1 undoubtedly show that the boundedness along with the scaling invariance gives rise to the shape $1/f^{\alpha(f)}$ not to $1/f$. The merit of

the shape $1/f^{\alpha(f)}$ is that it brings about finite variance of the fluctuations as proven in section *Variance. Setting of k* in Chapter 1. On the contrary, the shape $1/f$ yields infinite variance! Besides, the shape $1/f^{\alpha(f)}$ does not depend either on the statistics of the time series or its length. Note, that the withdrawal of the boundedness concept yields strong dependence of the power spectrum shape on the statistics of the time series. So, no universal shape would be possible! Alongside, if the variance of the noise is bounded, no presence of an irrational frequency in the discrete part would be possible! Hence, our theory takes away the major enigmas of the phenomenon $1/f$

Figure 2. A) Second time series B) The power spectrum of the time series in 2a

A)

B)

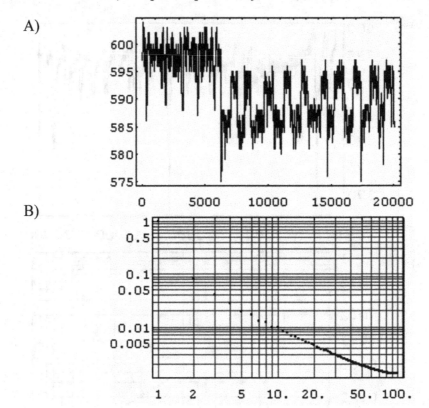

Variations of the Embedding Dimension

noise. That is why the establishing of non-constant exponent $\alpha(f)$ makes our arguments most persuasive. Our experiment on oxidation of $HCOOH$ is the first step in this direction. Let us just point out that the lack of data on linearity of $\alpha(f)$ is easily understandable: recall Table 1 where some values of κ are calculated. Since κ is vanishingly small, it is very difficult to be established if not looked for it purposely. Let us point out that we are the first one who have introduced and have been studying boundedness.

Summarizing, we assert that our concept successfully resolves the major mysteries of the $1/f$ noise. Then, the rich variety of systems that exhibit $1/f$ noise behavior strongly supports the advantage and ubiquity of the concept of boundedness.

The second aspect of the boundedness conjecture is the dynamical boundedness. The assertion is that the rate of development of fluctuations is bounded as a result of suggestion that the exchange of energy/matter with environment is kept permanently finite for every process in the Nature. Furthermore, in section *Boundedness of the Rates. Embedding dimension* in Chapter 1 we have developed an explicit relation between the dynamical boundedness and the embedding dimension: it varies with the length of the time series because the current value of the embedding dimension is set on the size of the largest in the time series fluctuation. On exerting the first "U-turn" the embedding dimension reaches its largest value and terminates its further variations on the increase of time series length. The examination of the experiment on the oxidation of $HCOOH$ shows that the current value of the embedding dimension manifests strong sensitivity to the fluctuation size for all 80 time series. The establishing of that persistent sensitvity is made by means of cutting the time series into pieces and monitoring the corresponding embedding dimension. It turns out that the latter permanently varies. In turn, this straightforwardly supports our finding that the current embedding dimension is related to the largest in the time series fluctuation size.

Homeostatic Patterns and Inter-Level Feedbacks

A very important part of our concept of boundedness is that of the hierarchical organization. We assert that the response of the system is sustained steady by means of self-organization into robust patterns and the hierarchical organization is substantiated by means of bounded noise. In turn, the inevitable presence of the inter-level feedback renders permanent deviations from the dynamical regime prescribed by the deterministic part of Equation (4.1) and Equation (7.13). As proved in Chapter 4, any difference $\left(\vec{x}\left(t\right) - \vec{x}_{\text{det}} \right)$ can effectively be presented as a solution of the corresponding equations-of-state at "shifted" control parameters. Note that at the same time the physical values of the control parameters are kept permanently fixed. The "shifting" causes immediate change either of the characteristics of the original dynamical regime or it even induces a bifurcation. It is obvious that at given parameter choice, an induced bifurcation needs development of a fluctuation of appropriate size. The induced bifurcation will contribute to the power spectrum as a discrete band. Note, however, that the induced bifurcation has temporary effect – it lasts until the fluctuation size is significant. Thus, by means of induced bifurcations we can trace the process

of spontaneous transition from one homeostatic pattern into another, (from one basin-of attraction into another), i.e. we can trace the process of spontaneous "writing of letters".

Let us now present an induced bifurcation that lasts until the fluctuation is significant. We took one particular time series recorded at the oxidation of $HCOOH$ and cut it into three successive parts of equal length. In Figure 3 those three parts in order of their succession are presented. In Figure 4, the corresponding power spectra are presented.

The examination of Figure 3 and Figure 4 apparently shows that the power spectrum of all three parts comprises both discrete and continuous band of the shape $1 \big/ f^{\alpha(f)}$ where $\alpha\left(f\right) \to 1$ at the low frequencies and grows linearly with the frequency. It is obvious that the presence of a large fluctuation (Figure 3b) strongly manipulates the amplitude of the discrete band in the corresponding power spectrum (Figure 4b): it is about 10 times smaller than that of the discrete bands in Figure 4a and Figure 4c. On the other hand, the period remains the same at all 3 power spectra. The great sensitivity of the amplitude of oscillations to the distance to the bifurcation point along with robustness of the period is a property genuine for a limit cycle. Further, the impact of that large fluctuation is temporary - it lasts as long as the fluctuation is essentially large. Indeed, as we already mentioned, the amplitude of the limit cycle at Figure 4a and Figure 4c is 10 times greater than that of the Figure 4b. Still, a closer look on Figure 4a and Figure 4c shows that there is a difference in the amplitudes of the discrete bands though not as pronounced as the one in Figure 4b. Therefore, we come to the conclusion that the effective shift of the control parameters is governed by the current largest fluctuation.

Further, it is important to stress that farthest left component in all 3 power spectra is an irrational one which once again confirms the boundedness of the noise and its relation with an inter-level feedback.

Figure 3. 3 successive parts of the time series of the oxidation of HCOOH

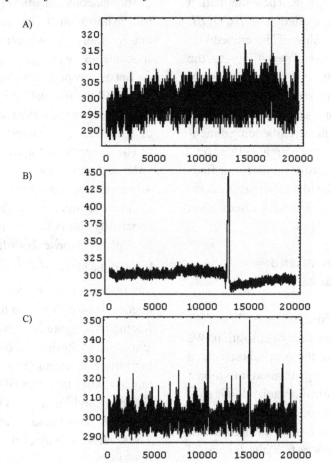

Thus, we conclude that the most important aspects of our theory are confirmed by the considered experiment. On the other hand, the same experiment cannot be explained in the frame of traditional chemical kinetics since it cannot explain either the persistent presence of a noise band in the power spectra or its coexistence with a discrete band along with the spontaneous development of noise-induced bifurcations.

SOCIAL SYSTEMS UNDER BOUNDEDNESS

In the previous section we have considered how our concept of boundedness is manifested in an experiment taken from one of the simplest types of complex systems. Now we will focus attention at the other end of the spectrum, on the most complex systems - social ones. The challenge consists of the fact that we, human beings, have the ability actively to change the environment where we live. That is why it is very important to have reliable instruments for predicting the effect of our intervention. The fundamental difficulty in this task is that, taken as a whole, our impact is 'non-linear' and 'non-homogeneous'. In turn, the effect of this impact makes any long-term prediction a hopeless task. Moreover, current approaches to the problem are grounded on the assumption that the hierarchy of the social systems is of a pyramidal type. This type of hierarchy, however, has the

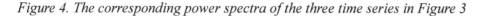

Figure 4. The corresponding power spectra of the three time series in Figure 3

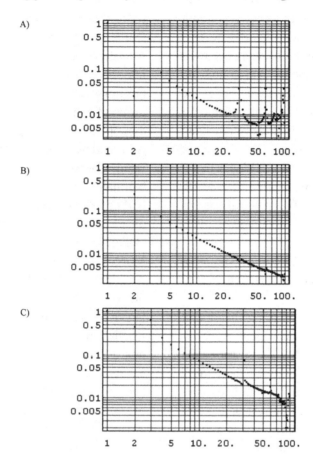

flaw to be extremely vulnerable to any uncertainty of the corresponding classification. That is why this approach requires very accurate preliminary information. Also, we have to take into account the fact that the information in social systems is fundamentally inaccurate because of the following reasons: (i) often we react as human beings: both emotionally and logically; (ii) as we have already established in the present chapter, the logic, based on answering binary questions is not enough for obtaining accurate knowledge; (iii) sometimes humans are reluctant to say or act reasonably because of a variety of reasons. In general, we can conclude that there are fundamental reasons for considering any preliminary information about social systems uncertain.

At this point the theory of boundedness could be of great help in explanation of the behavior of social systems. First of all, the hierarchical self-organization under boundedness selects only stable solutions. This implies that it either preserves current homeostasis, or drives its changes towards another stable pattern. The question is whether we can 'foresee' these changes if the collected information is uncertain? Our answer is affirmative, yes, we can 'foresee' even if the available information is uncertain. The major implement for solving this task is grounded on the assumption to regard any gathered information as a BIS. Then, on the grounds of the property about additive separation of the power spectrum of each BIS to a discrete and continuous band, we can make the following: by means of applying the operation of coarse-graining to a BIS several times followed by separation of the corresponding power spectrum, we can discriminate between a steady and unsteady behavior. A steady behavior is realized when the corresponding discrete band in the power spectrum remains intact under the coarse-graining; accordingly, unsteady behavior is realized when the corresponding discrete band changes under coarse-graining.

The advantage of the above criterion is that it operates with uncertain information. It is worth noting that the application of coarse-graining is a genuine 'simulation' and does not involve gathering additional information.

Next we present argumentation in favor of adopting non-reductionist approach to social systems.

Social behavior considered as a realization of numerous strategies by different agents provides one of the greatest mysteries in the world. Some local events bound to instability by the current religious, historical, cultural, etc. milieus suddenly become definitive in the rearrangements of the social priorities that take place in a given community. Rationality and various reductionist social models based on automatic interplay of social groups, modes of production, and functions of social structures do not provide sufficient answers that can explain these sudden social changes.

We will try to outline the paths leading to such 'non-linear' behavior of sudden changes of the public opinion. Dominant systems of community values shape to a considerable degree individual opinions and behavior. Opinion polls, however, show diversified responses and their results remain away from a simple sum of individual answers. Let me start with some social preferences that modify personal opinions. We are very positive on certain topics, beliefs, and predispositions, but remain hesitant in some others. For example most people believe that there exist alien forms of life in the Universe. But to the question: 'Are aliens friendly to us?' public opinion splits approximately in a half. For the purpose of this consideration it is enough to know that we have definite opinion in some issues, while others make us feel uncertain. Our aim is to describe the general path that leads from destabilization of positive personal opinions into social instability that rearranges some priorities in a given community. Such factors may have different nature, but their essential quality is that they operate locally or globally. The local behavior is formed by the closest circle of people we love, family, friends that constitute our 'neighborhood'. Normally, we share same opinions. Yet, the neighborhood may be split on important issues and sends disorienting signals to its members. Opinions can change from positive to negative and vice versa. Moreover, the uncertainty is "contagious". Our "neighbors "infect" their other "neighbors" and so the process of destabilization spreads throughout larger parts of the society. Now the question is whether the destabilization is limited to scattered non-overlapping areas or it is spread over the entire society. The importance of this question becomes evident from the following: in the case of isolated "islands" of uncertainty embedded in a "sea" of certainty, the dominant opinion comes from the "sea". Thus, it is insensitive to the current events and does not change significantly with the course of time. If, however, the uncertainty prevails, i.e. it is above the percolation threshold, the vote exhibits minor random variations around the "neutral" line - 50:50. This makes the social support "blind" - the winner is elected by a "close margin' or sometimes is unpredictable as "a tossing of a coin".

Our experience teaches us that the psychological and emotional arguments make us more confident in ourselves than any logic. However, isolated and non-correlated episodes of individual stabilization do not help the global one. The second major source of influence on the individual opinion, the global one, comes from the viewpoint of our "idols", movie stars, journalists, favorite shows, delivered instantly to every individual by the Information and Communication Technologies (radio, TV, Internet). Thus, if some local event of a great psychological and/or social impact is presented as emblematic and is strongly supported by the celebrities, its effect stabilizes simultaneously the public opinion through the stabilization of each individual. The hallmark of the stabilization is the coherence - all the individuals vote for the same alternative, i.e. alternative "chosen" by the effect of the emblematic "hot case". Thus, the society supports that decision-making strategy whose top priority is situated at best to the current emblematic event(s).

There are limitations imposed on the impact of 'exemplary' behavior to environment. In case when the exchange of information and accumulation of new knowledge grows fast the reaction of the environment becomes more selective, smart, and unpredictable relative to the expected outcomes. As a result of 'overloading' with 'exemplary cases' the system as a whole may or may not behave in an unpredictable way. The stabilization of such system will depend to a larger extent on the ability of the system's 'moderators' to balance the input of adequate amount of new information so that any transitional states to remain bounded.

The role of the psychological aspect of the local events is well known and is widely exploited nowadays. But it has been exploited before the TV time as well. Let us go back to the medieval times, to the discovery of America by C. Columbus. The new continent was named not after him but after the name of Amerigo Vespucci, the man who has become popular because he proclaimed the discovered by C. Columbus land "The New World". The magic of the collocation that creates the phrase "The New World" gave a strong hope for a new life for the old world though the people knew a little or nothing about it.

CONCLUSION

We have started this book by posing the question what makes both intelligent and non-intelligent systems exhibit certain common properties? Is this an indication that the intelligence is implemented by natural processes and could it be exerted autonomously, i.e. not being a subject to a hand-craft manipulation? The problem is very important since it is well known that the computers and other intelligent devices are not autonomous in creating and comprehending information. Their non-autonomity is grounded on the fact that neither of them could discriminate between a random sequence of symbols and a meaningful one. That is why we assert that the greatest achievement of our theory of boundedness is that it gives credible grounds for autonomous creation and autonomous comprehending of information. This is rendered possible by means of organizing the functionality of a complex system so that it to select automatically the corresponding semantics through performance of specific engines each associated with a specific semantic unit. Thus, we open the door to creating a general strategy for building a functional circuit able to autonomous creation and comprehending information. Its major advantage would be that such circuit would operate so that to sustain balance with its environment.

REFERENCES

Bennett, C. H. (2003). Notes on Landauer principle, reversible computation and Maxwell Demon. *Studies in History and Philosophy of Modern Physics*, *34*, 501–510. doi:10.1016/S1355-2198(03)00039-X

Koleva, M. K., Elyias, A. E., & Petrov, L. A. (2000). Fractal power spectrum at catalytic oxidation of HCOOH over supported Pd Catalyst . In Russo, N., & Salahub, D. R. (Eds.), *Metal-ligand interactions in chemistry, physics and biology, NATO ASI Series C* (*Vol. 546*, pp. 353–369). Dordrecht, The Netherlands: Kluwer Academic Publishers. doi:10.1007/978-94-011-4245-8_15

Norton, J. D. (2005). Eaters of the lotus: Landauer's principle and the return of Maxwell Demon. *Studies in History and Philosophy of Modern Physics, 36*, 375–411. doi:10.1016/j.shpsb.2004.12.002

Parker, M. C., & Walker, S. D. (2004). Information transfer and Landauer principle. *Optics Communications, 229*, 23–27. doi:10.1016/j.optcom.2003.10.019

Plenio, M. B., & Vitelli, V. (2001). The physics of forgetting: Landauer erasure principle and information theory. *Contemporary Physics, 42*, 25–60. doi:10.1080/00107510010018916

Shenker, O. R. (2000). *Logic and entropy.* Retrieved from http://philsci-archive.pitt.edu/archive/00000115/

Woodward, J. (2003). *Making things happen: A theory of causal explanation.* Oxford, UK: Oxford University Press.

Zander, C., Plastino, A. R., Plastino, A. P., Casas, M., & Curilef, S. (2009). Landauer`s principle and divergenless dynamical systems. *Entropy, 11*, 586–597. doi:10.3390/e11040586

Epilogue

BOUNDEDNESS AND SELF-ORGANIZED SEMANTICS IN A NUTSHELL

This book started by posing the question why a great amount of both non-intelligent and intelligent complex systems share the same properties and what makes a system to respond in an intelligent way. What we have learnt from the book is that a range of common properties of both intelligent and non-intelligent systems is necessary for sustaining the stability of a system. Further, the stability is developed through hierarchical self-organization which, to the most surprise, takes a new role: the role of implement of a semantic-like response. These premises, as we have seen, define an operational protocol that allows developing a systematic general theory about transformation of a system of constituents into a system of specifiable agents which respond to external stimuli in an intelligent-like manner. The proposed approach develops both the physics and the information theory and puts the relation between them on novel grounds by establishing the foundation for realization of a functional circuit capable to autonomous organization of information in a hierarchy of semantic structures along with its autonomous comprehending and autonomous creating of information.

The book has demonstrated that the key advantages of this strategy are:

- The boundedness sets the homeostasis to be a property of each hierarchical level starting with the quantum-mechanical one. Alongside, it makes available its characteristics to take metric-free forms which provides covariance of their robustness to small perturbations (provides their independence from the choice of a reference frame).
- The exclusive property of boundedness to appear both as a grounding principle for every given hierarchical level and as an emergent property at the next one opens the door to versatility and scalability of the hierarchical order through its non-extensive super-structuring.
- The general way of 'building' a hierarchy is through morphogenesis considered in the setting of the developed by us concept of boundedness. Its major advantage is that it provides a sustainable evolution of a 'kind' by means of diversifying its mutations.
- The semantics appears as an autonomous property of self-organization contrary to the traditional information theory where the process of distinguishing of a message from a random sequence is subject to an "external mind."
- The information is organized in semantic order in a non-extensive manner. The effect of that non-extensivity is best illustrated by the rate of its extent – it is justified by the availability the next and the previous unit (letter, word, sentence…) to be "foreseen" on the grounds of the knowledge about the current unit only, i.e. such circuit performs as an "Oracle." To compare, according to the

traditional information theory each letter can follow or precede any other one from the alphabet and thus it is impossible not only "foreseeing" the next letter, but it is impossible to judge *a priori* whether a given sequence is a message (i.e. it comprises information) or it is a random sequence of symbols.

- The boundedness sets an exclusive two-fold representation of a semantic unit: as a specific sequence of letters and as performance of a specific "engine." The latter renders a novel approach to the causality that acts in the hierarchical order - the functional irreversibility of any Carnot engine (in one direction works as a heat machine and in the opposite it performs as a refrigerator) provides a way of implementing logical irreversibility as a generic property of any semantic unit.

- The exclusive two-fold representation of a semantic unit: as a specific sequence of letters and as performance of a specific engine serves as grounds for building a multi-layer hierarchy of semantic structures. It is established that in this setting the semantics admits both non-extensivity, permutation sensitivity, and Zipf's law. This approach proves the governing role the Second Law plays in the operation of the "engines." Yet, the fundamental breakthrough is that the Second Law is considered as putting ban over any realization of *Perpetuum mobile* at the expense of releasing it from the requirement for maximization of the entropy. In turn, the latter renders the widest ubiquity of the Second Law: it adds to its realm the entire variety of complex systems including systems that exhibit different forms of macroscopic self-organization (e.g. pattern formation and morphology), i.e. systems whose common property is violation of the entropy maximization as condition for thermodynamical equilibrium.

- The information is defined in a metric-free form, irrational numbers included, the major property of which is the robustness to small perturbations. This opens the door to "abstract" way of defining objects: both "red" and "green" apple belong to the set of "apples," yet "red" and "green" are metric properties while "apple" is an invariant for the entire class of "apples" that possesses metric-free property.

Compilation of References

Bak, P., & Sneppen, K. (1993). Punctuated equilibrium and criticality in a simple model of evolution. *Physical Review Letters, 71*(24), 4083–4086. doi:10.1103/PhysRevLett.71.4083

Bak, P., Tang, C., & Wiesenfeld, K. (1987). Self-organized criticality: An explanation of 1/f noise. *Physical Review Letters, 59*(4), 381–384. doi:10.1103/PhysRevLett.59.381

Bennett, C. H. (2003). Notes on Landauer principle, reversible computation and Maxwell Demon. *Studies in History and Philosophy of Modern Physics, 34*, 501–510. doi:10.1016/S1355-2198(03)00039-X

Feller, W. (1970). *An introduction to probability theory and its applications.* New-York, NY: John Wiley & Sons. doi:10.1063/1.3062516

Gardiner, C. W. (1985). *Handbook of stochastic methods for physics, chemistry and the natural science* (Haken, H., [Ed.] *Vol. 13*). Berlin, Germany: Springer Series in Synergetics. doi:10.2307/2531274

Hughes, C. P., & Nikeghbali, A. (2008). The zeroes of random polynomials cluster uniformly near the unit circle. *Compositio Mathematica, 144*, 734-746. Retrieved from http://arXiv.org/abs/math.CV/0406376

Kauffman, S. (1995). *At home in the universe: The search for laws of self-organization and complexity.* New York, NY: Oxford University Press.

Koleva, M. K. (2005). *Fluctuations and long-term stability: From coherence to chaos.* Retrieved from http://arxiv.org/abs/physics/0512078

Koleva, M. K., & Covachev, V. (2001). Common and different features between the behavior of the chaotic dynamical systems and the 1/f^alpha-type noise. *Fluctuation and Noise Letters, 1*(2), R131-R149. arXiv:cond-mat/0309418

Koleva, M. K., Elyias, A. E., & Petrov, L. A. (2000). Fractal power spectrum at catalytic oxidation of HCOOH over supported Pd Catalyst. In Russo, N., & Salahub, D. R. (Eds.), *Metal-ligand interactions in chemistry, physics and biology, NATO ASI Series C* (*Vol. 546*, pp. 353–369). Dordrecht, The Netherlands: Kluwer Academic Publishers. doi:10.1007/978-94-011-4245-8_15

Norton, J. D. (2005). Eaters of the lotus: Landauer's principle and the return of Maxwell Demon. *Studies in History and Philosophy of Modern Physics, 36*, 375–411. doi:10.1016/j.shpsb.2004.12.002

Parker, M. C., & Walker, S. D. (2004). Information transfer and Landauer principle. *Optics Communications, 229*, 23–27. doi:10.1016/j.optcom.2003.10.019

Plenio, M. B., & Vitelli, V. (2001). The physics of forgetting: Landauer erasure principle and information theory. *Contemporary Physics, 42*, 25–60. doi:10.1080/00107510010018916

Shenker, O. R. (2000). *Logic and entropy.* Retrieved from http://philsci-archive.pitt.edu/archive/00000115/

Turing, A. (1952). The chemical basis of morphogenesis. *Philosophical Transactions of the Royal Society B, 237*, 37–72. doi:10.1098/rstb.1952.0012

Woodward, J. (2003). *Making things happen: A theory of causal explanation.* Oxford, UK: Oxford University Press.

Zander, C., Plastino, A. R., Plastino, A. P., Casas, M., & Curilef, S. (2009). Landauer`s principle and divergenless dynamical systems. *Entropy, 11*, 586–597. doi:10.3390/e11040586

About the Author

Maria K. Koleva, PhD, has been Associate Professor in the Institute of Catalysis, Bulgarian Academy of Sciences, Sofia, Bulgaria since 2007. In the recent decade, her main interest is focused on the systematic development of the theory of boundedness. The approach sets the leading role of the principle of boundedness for the hierarchical self-organization of matter, energy and information. The hypothesis of boundedness consists of 1) a mild assumption of boundedness on the local (spatial and temporal) accumulation of matter/energy at any level of matter organization, and 2) boundedness of the rate of exchange of such an accumulation with the environment. Exclusive property of the proposed approach is the intelligent-like way of self-sustaining the functionality of the multi-layer hierarchy by means of governing the intra- and inter-level homeostasis through rules that are organized in semantic-like manner. The major results of the approach were reported at a talk given at Niels Bohr Institute in 2009.

Index